AF327178

PHASE STABILITY IN
HIGH TEMPERATURE ALLOYS

Proceedings of a course held at the

JOINT RESEARCH CENTRE
of the
COMMISSION OF THE EUROPEAN COMMUNITIES
Petten Establishment, The Netherlands

and organised by the

CEC High Temperature Materials
Information Centre,
Petten, The Netherlands

PHASE STABILITY IN HIGH TEMPERATURE ALLOYS

Edited by

V. GUTTMANN

Joint Research Centre, Petten Establishment, The Netherlands

APPLIED SCIENCE PUBLISHERS LTD
LONDON

APPLIED SCIENCE PUBLISHERS LTD
RIPPLE ROAD, BARKING, ESSEX, ENGLAND

British Library Cataloguing in Publication Data

Phase stability in high temperature alloys.
 1. Heat resistant alloys
 2. Phase rule and equilibrium
 I. Guttmann, Viktor
 669′.94   TN700

ISBN   0-85334-946-0

WITH 4 TABLES AND 89 ILLUSTRATIONS

© ECSC, EEC, EAEC, BRUSSELS AND LUXEMBOURG, 1981

Publication arrangements by: Commission of the European Communities, Directorate-General for the Information Market and Innovation, Luxembourg

EUR 7106 EN

LEGAL NOTICE

Neither the Commission of the European Communities nor any person acting on behalf of the Commission is responsible for the use which might be made of the following information.

Printed in Great Britain by Galliard (Printers) Ltd, Great Yarmouth

Preface

This book contains the lectures presented at a course on 'Phase Stability in High Temperature Alloys', organised by the Commission of the European Communities, Joint Research Centre, Petten Establishment.

Phase stability becomes one of the most important metallurgical problems when metals and alloys are used at elevated temperatures. The processes which determine the microstructure, in particular solidification and thermomechanical treatments, are of a short-term nature and generally metastable conditions are obtained. The use in service of such materials is frequently accompanied by unavoidable structural changes, such as phase transformation, particle ripening, grain growth, etc., which can lead to serious changes in the material properties.

The lectures were intended to present a summary of relevant theoretical and practical aspects of metal structures with emphasis on their possible changes during service. Each chapter provides the reader with an overview of a specific subject. The first two chapters are concerned with fundamental thermodynamic aspects and with computational techniques of phase diagrams, respectively. The third chapter deals with the interaction between lattice defects and particle generation. The following contribution is devoted to the influence of service conditions on phase development and transformation. In the final chapter the problem of phase instability is considered with specific emphasis on the material application under engineering conditions.

V. GUTTMANN

Contents

1

Fundamental Thermodynamic Aspects of Phase Diagrams*

W. PITSCH

*Max Planck Institut für Eisenforschung GmbH,
Düsseldorf, FRG*

ABSTRACT

A large variety of structures can be produced in metallic materials by means of phase-transformations. The stability of these structures is governed by the thermodynamics of the particular metallic alloy. Therefore the knowledge of the thermodynamic functions of any phase, which may occur in a certain alloy, is the pre-requisite for any estimate of the mutual stability of different structures. The formalisms of these thermodynamics will be summarised in the following with particular emphasis on iron and its alloys. In addition, by means of atomistic models the physical origins of stability will be illustrated and a better understanding obtained.

1. PURE METALS

1.1 Introductory Remarks

The thermodynamic rules indicate how stable at a certain temperature one phase is in comparison with another phase, e.g. the solid phase of a pure metal with a certain lattice structure compared with the melt.† This comparison is made with the aid of a temperature dependent function, the Gibbs energy G, which is composed of the enthalpy H and the entropy S:

$$G(T) = H(T) - T.S(T) \tag{1}$$

* Revised translation of 'Thermodynamik des Eisens und seiner Legierungen' in *Grundlagen der Wärmebehandlung von Stahl* (Ed. W. Pitsch), Verlag Stahleisen mbH, Düsseldorf (1976).
† For the sake of simplicity the dependence of phase-stability on pressure will not be treated.

1

In order to make these comparisons, it is desirable to know $G(T)$ for every phase φ at every temperature T. The criterion for stability is: the more negative the function G of a certain phase is, the more stable is that phase. Or in other words: with respect to every other phase which has a more negative Gibbs energy G than the considered phase there is a tendency for a phase-transformation.

The Gibbs energy $G(T)$ is obtained from measurements of the specific heat $c_p(T)$ at constant pressure. It is

$$dH = c_p(T) \cdot dT \qquad \text{and} \qquad dS = \frac{c_p(T)}{T} \cdot dT \tag{2}$$

Therefore it follows that

$$H(T) = H(0) + \int_0^T c_p(\vartheta)\, d\vartheta \tag{3}$$

and

$$S(T) = S(0) + \int_0^T \frac{c_p(\vartheta)}{\vartheta}\, d\vartheta \tag{4}$$

and finally

$$G(T) = H(0) - T \cdot S(0) - \int_0^T \left[\frac{T}{\vartheta} - 1 \right] c_p(\vartheta)\, d\vartheta \tag{5}$$

The entropy can also be deduced (as well as from eqn (4)) from the gradient of the $G(T)$ curve since differentiating eqn (1) and using eqn (2) gives:

$$\frac{dG(T)}{dT} = -S(T) \tag{6}$$

It follows from eqn (5) that the function $G(T)$ for a phase φ is fully determined if the entities $H(0)$, $S(0)$ and $c_p(T)$ are known. However, frequently the determination of $c_p(T)$ is not possible by direct measurements. Then suitable estimates have to be obtained by extrapolations and/or interpolations of the measurements; physical models are important guides for these procedures. Such models are available for the different physical effects which contribute to the $c_p(T)$ curve. These effects are connected with the energy absorptions of the atoms (lattice vibrations and thermal expansion), of the free electrons, and for iron and its alloys of the localised energy states, of the 3d electrons (magnetic effects).

For qualitative considerations of phase stabilities it is frequently desirable to have a certain idea of the meanings of $H(T)$ and $S(T)$. Such a conception is obtained, if one takes $H(0)$ as the cohesion energy of the atoms at $T = 0$. Then it is obvious that $H(0)$ is the more favourable, i.e. the more negative, the stronger the atoms are bounded.[1] With increasing temperature the atom bonds are weakened and accordingly more unfavourable states of the atoms and electrons are occupied and thermal energy is absorbed. The function $H(T)$ increases with temperature, in accordance with eqn (3), on to more positive values and its contribution to the thermodynamic stability becomes less.

The entropy can be conceived as a measure of the uncertainty of our knowledge about the state of the phase.[1] This uncertainty increases as the number of ways in which each state of a phase can be composed of the different states of atoms and electrons increases. At $T = 0$ a crystalline phase is—if it is not degenerated—uniquely defined and therefore, according to the Nernst law of heat, the zero point entropy is $S(0) = 0$. With increasing temperature the atoms and electrons are transformed into states of higher energy. There is an increasing variety of energetically equivalent modes for this. Therefore the state of the crystal can no longer be uniquely defined. According to eqn (4) this is accompanied by an increase in the values of the entropy. Hence the entropy contribution to the thermodynamical stability increases with increasing temperature according to eqn (1). In fact according to eqn (5) the influence of the entropy actually predominates: the Gibbs energy always decreases to more negative values with increasing temperatures.

1.2 Non-magnetic Metals

The relationships between $c_p(T)$ and the thermodynamic functions G and H as described by eqns (3) and (5) are presented schematically in Fig. 1 for the most simple case of a pure non-magnetic metal. The $G(T)$ curve starts at $T = 0$ with the value $H(0)$ and a gradient of $S(0) = 0$. The $H(T)$ curve increases, and the $G(T)$ curve decreases, with increasing temperature.

The relative stability of two phases $\varphi = \alpha, \beta$ in a pure element is determined by the difference

$$\Delta G^{\alpha/\beta}(T) = G^{\beta}(T) - G^{\alpha}(T) \tag{7}$$

In order to evaluate this difference, the $c_p(T)$ curves of the two phases have to be determined either by measurements or by estimates.[2] The zero point entropies are set equal to zero for non-degenerate states. The zero point enthalpies $H^{\alpha}(0)$ and $H^{\beta}(0)$ are at first unknown. However their relative

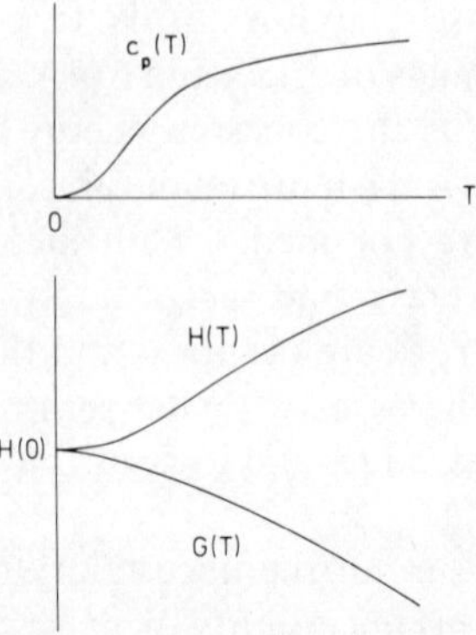

Fig. 1.   Specific heat c_p, enthalpy H and Gibbs energy G of a pure, non-magnetic metal, dependent on temperature T (schematic).

Fig. 2.   Enthalpies and Gibbs energies of two phases α, β (schematic).

positions, that is the difference $\Delta H^{\alpha/\beta}(0)$, is deduced from the equilibrium condition at the known transition temperature T_0:

$$\Delta G^{\alpha/\beta}(T_0) = 0 \tag{8}$$

With these data the relative positions of the curves $G^{\alpha}(T)$ and $G^{\beta}(T)$ can be fully determined (Fig. 2). From this figure it follows that at low temperatures, phase α, with the more negative $H(0)$, is more stable than phase β. Physically this means, that the atoms in phase α are more strongly bound and therefore less movable than in phase β. Therefore during heating, phase β absorbs more heat than phase α; accordingly $c_p^{\beta}(T)$ and $H^{\beta}(T)$ increase more strongly than $c_p^{\alpha}(T)$ and $H^{\alpha}(T)$. Simultaneously $G^{\beta}(T)$ decreases more strongly than $G^{\alpha}(T)$. The Gibbs energy curves cross each other at the equilibrium temperature T_0. Above T_0 the β-phase is more stable than the α-phase. For illustration purposes the α-phase can be imagined as a solid phase and the β-phase as the melt.

The transformation enthalpy $H^{\beta}(T_0) - H^{\alpha}(T_0)$ is frequently known by direct measurements. From eqns (1) and (8) the change of entropy at T_0 is as follows:

$$S^{\beta}(T_0) - S^{\alpha}(T_0) = \frac{H^{\beta}(T_0) - H^{\alpha}(T_0)}{T_0} \tag{9}$$

It is known from experience that for non-magnetic metals above $T = 0$ the differences in Gibbs energies are practically linearly dependent upon the

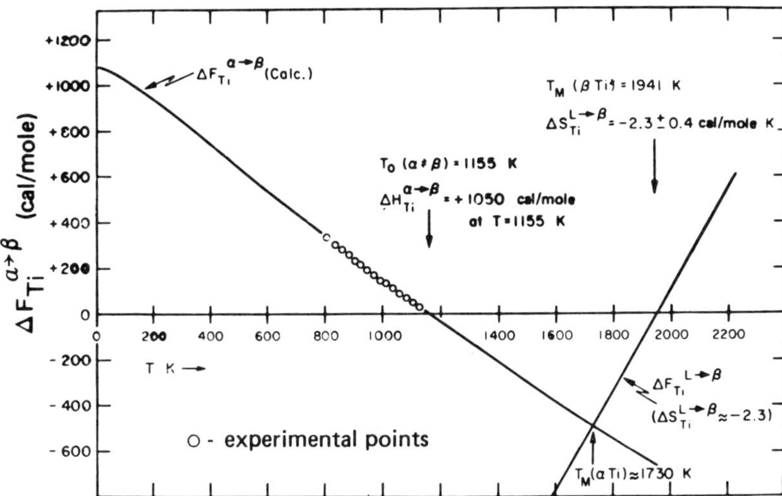

FIG. 3.   Difference in Gibbs energy (here indicated as ΔF) for the hexagonal close-packed α-phase, the cubic body centred β-phase and the liquid, L, phase of titanium.[3]

temperature T. An example is given in Fig. 3, where the difference in the Gibbs energy of the hexagonal close-packed α-phase and the cubic body centred β-phase of pure titanium is presented.[3] In addition a similar curve is included for the β-phase and the melt, L, phase.

1.3 Iron

The thermodynamic stability of non-magnetic metals is mainly determined by the states of energy of the atoms and free electrons. These states will be indicated in the following by A and E. For magnetic metals such as iron, there is an additional contribution from magnetic effects (M) which strongly determines the polymorphism of this element. As an example Fig. 4 presents the measured $c_p(T)$ curve of phase α-Fe (and δ-Fe) with a cubic body centred lattice.[4] The transition from magnetic order to magnetic disorder with increasing temperature is indicated by the particular absorption of heat close to the Curie point, $T_c = 1042\,°C$. Without this magnetic effect the $c_p(T)$ curve of α-Fe would be equal to curve B in Fig. 4. The total absorption of energy due to the magnetic effect is equal to the area between the measured curve and curve B in Fig. 4. This energy, $[H^{A+E+M}(0) - H^{A+E}(0)]$, is equal to almost 8250 J/mol Fe-atoms for α-Fe.

The $c_p(T)$ curve B in Fig. 4 corresponds to the non-magnetic energy absorption of α-Fe. With the aid of physical models this curve can be

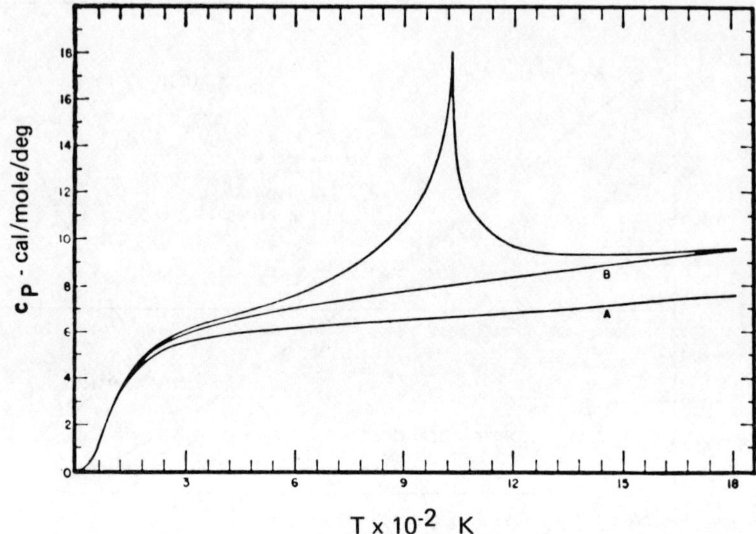

FIG. 4.   Specific heat of α (bcc) Fe. Curve A is the calculated lattice specific heat for a Debye $\theta = 420\,^\circ$K. Curve B includes the electronic specific heat.[4]

further separated into; (i) the contribution of the atoms which is equal to the lattice vibrations and the thermal expansion (curve A), and (ii) the contribution of the conducting electrons which is equal to the difference between curves B and A.

The Gibbs energies which follow from the $c_p(T)$ curves in Fig. 4 are discussed in Fig. 5: the stability of a crystal without magnetic moments are presented by the enthalpy and Gibbs energy curves $H^{A+E}(T)$ and $G^{A+E}(T)$ which follow from curve B in Fig. 4. The additional effect of the magnetic

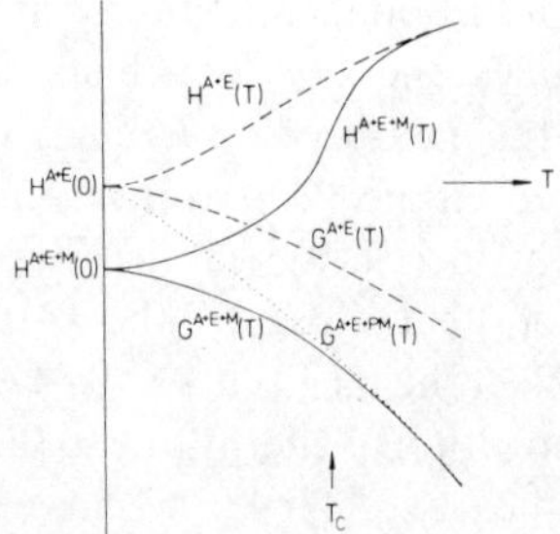

FIG. 5.   Enthalpies and Gibbs energies of a phase, which is non-magnetic (A + E), magnetic (A + E + M) or at all temperatures paramagnetic (A + E + PM) (schematic).

contribution is most pronounced at $T = 0$, where due to the additional bounding of the (parallel) magnetic moments the zero point enthalpy is shifted to a more negative value. Correspondingly the Gibbs energy of a magnetic phase at $T = 0$ is also shifted by

$$G^{A+E+M}(0) - G^{A+E}(0) = H^{A+E+M}(0) - H^{A+E}(0)$$

$$= \int_0^\infty [c_p^{A+E+M}(\vartheta) - c_p^{A+E}(\vartheta)]\,d\vartheta \qquad (10)$$

With increasing temperature the transition from magnetic order to magnetic disorder causes an additional absorption of energy. Therefore, the enthalpy curve of a magnetic phase approximates that of the non-magnetic contributions, with

$$H^{A+E+M}(T) = H^{A+E}(T) - H^{M}(T) \qquad (11)$$

with

$$H^{M}(T) = \int_T^\infty [c_p^{A+E+M}(\vartheta) - c_p^{A+E}(\vartheta)]\,d\vartheta \qquad (12)$$

Using the total specific heat curve in Fig. 4 the Gibbs energy curve $G^{A+E+M}(T)$ follows from eqn (5) starting at $T = 0$ with the value $H^{A+E+M}(0)$ and with a gradient given by the zero point entropy $S^{A+E+M}(0) = 0$. Due to the magnetic bonds this Gibbs energy curve is more negative than the Gibbs energy curve $G^{A+E}(T)$ of the non-magnetic contributions.

For illustration purposes the Gibbs energy $G^{A+E+PM}(T)$ is also indicated in Fig. 5 and this corresponds to a hypothetical phase which is paramagnetic (PM) at all temperatures. At high temperature it is natural that $G^{A+E+PM}(T)$ and $G^{A+E+M}(T)$ are practically identical. Since however with decreasing temperatures no magnetical alignment is allowed to occur there is no magnetical enthalpy contribution to the phase stability. From this it follows that compared to the magnetic state, $A + E + M$, there is less stability which means a more positive Gibbs energy $G^{A+E+PM}(T)$. However, since at all temperatures, including $T = 0$, the random orientations of the magnetic moments cause a degenerated state, there is a constant contribution of magnetic entropy to phase stability at all temperatures, which is equal to

$$S^{M}(0) = \int_0^\infty \frac{c_p^{A+E+M}(\vartheta) - c_p^{A+E}(\vartheta)}{\vartheta}\,d\vartheta \qquad (13)$$

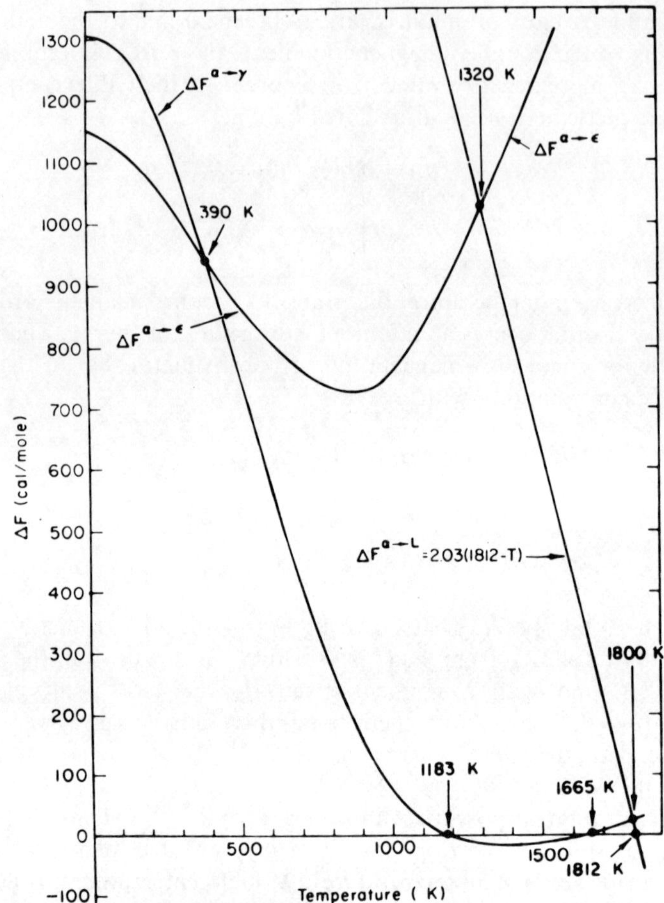

FIG. 6.   Differences of Gibbs energies (here indicated as ΔF) of the bcc (α), fcc (γ), hcp (ε) and liquid (L) phases of iron.[5]

Therefore, in comparison with the non-magnetic state, A + E, the state, A + E + PM, is more stable. For the degenerate state, A + E + PM, the zero point entropy is $S^M(0) \neq 0$.

The described magnetic effects occur mainly in the cubic body centred phases α-Fe and δ-Fe. They occur to a less extent in the cubic face centred phase γ-Fe and even less in the hexagonal close-packed phase ε-Fe and in the liquid phase L-Fe. The differences in the Gibbs energies of the different phases of iron are therefore more complicated than those of non-magnetic metals. One of the most recent quantitative determinations[5] is presented in

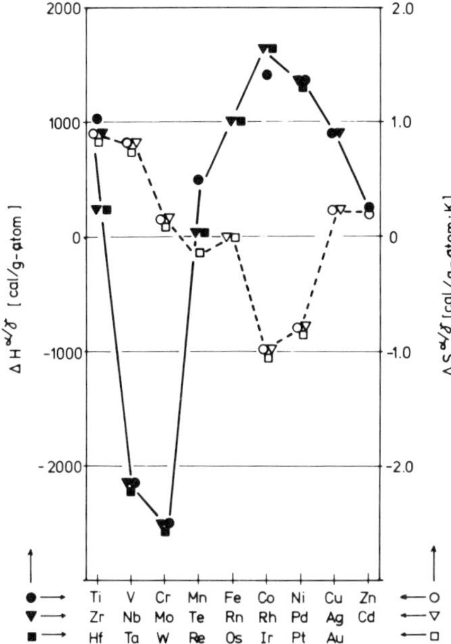

FIG. 7.   Differences in enthalpy (filled symbols) and in entropy (unfilled symbols) for bcc (α) and fcc (γ) phases in pure metals; magnetic effects are not accounted for.[5]

Fig. 6. Considering the curve for the α-phase and γ-phase it is seen, that with decreasing temperature, at first—as is expected—the less close-packed α-phase becomes less stable than the more close-packed γ-phase. At about 1400 K however this tendency reverses, since with the onset of the strong magnetic bounding effects in the α-phase, this phase becomes more stabilised. At 1400 K the γ-phase is more stable than the α-phase. Therefore it follows that two transition points occur; at 1665 K (A4) and at 1183 K (A3).

In a similar way the differences in the Gibbs energies have been estimated for many metals and different phases.[5,6] For non-magnetic metals the temperature dependence of these differences could be approximated satisfactorily by a simple linear function (see also Fig. 3):

$$\Delta G^{\alpha/\beta}(T) = \Delta H^{\alpha/\beta}(T_0) - T . \Delta S^{\alpha/\beta}(T_0) \tag{14}$$

The constants in the preceding equation, the so-called 'stability parameters', have been evaluated for many metals. An example, taken from reference 5 is presented in Fig. 7 for the cubic body centred and cubic face centred phases α and γ. For the magnetic metals Cr, Mn, Fe, Ni and Co, which are also included in Fig. 7, the data indicate the contributions which are solely due to the atoms and conducting electrons.

2.   SUBSTITUTIONAL BINARY SOLID SOLUTIONS

2.1 Gibbs Energies

The Gibbs energies of alloys are developed on the basis of the thermodynamic functions of the pure elements. If these Gibbs energies are known for all phases, alloy compositions and temperatures (at constant pressure), then all states of stable and metastable equilibria can be treated. The stable equilibria are presented in the well-known form of phase diagrams. Metastable equilibria frequently depend on certain additional conditions, e.g. the condition that the nucleation of the most stable phase is constrained, or that no long range diffusion and therefore no decomposition can take place in the alloy. The latter condition is fulfilled at low temperatures, where transformations are martensitic.[7]

Unfortunately, in general it is not possible to deduce the Gibbs energies of solid solutions, by means of eqn (5), from measurements and estimates of the specific heat. Instead analytical formulae have been modelled for the Gibbs energies and the parameters, incorporated into these formulae, are determined by means of thermodynamic measurements and/or by application of these formulae to known phase equilibria.[8,9] This procedure will be discussed below.

The composition of substitutional solid solutions AB is most conveniently described by the atom fractions x_A and x_B with $x_A + x_B = 1$. Then the following expression for the Gibbs energy of a phase φ has been found to be the most convenient:

$$G^{\varphi}(x_A T) = x_A{}^0 G_A^{\varphi}(T) + x_B{}^0 G_B^{\varphi}(T) + RT\{x_A \ln x_A + x_B \ln x_B\} + {}^E G^{\varphi}(x_A, T)$$

$$(15)$$

In the first and second term of the expression the Gibbs energies of the pure components A and B are added accordingly to their weights given by x_A and x_B. The third term is the entropy of mixing which corresponds to a random distribution of the atoms A and B on to the lattice sites. The last term corrects any deviation between the first three terms and the true value of G^{φ}. The determination of this term, the so-called excess Gibbs energy, is achieved in the same way as the determination of the Gibbs energy G^{φ}.

2.2 Equilibrium Conditions

If no decomposition occurs in the alloy, two phases $\varphi = \alpha, \beta$ are in metastable equilibrium with each other at a temperature T_0 which is given by

$$\Delta G^{\alpha/\beta}(x_A, T_0) = G^\beta(x_A, T_0) - G^\alpha(x_A, T_0) = 0 \tag{16}$$

From this expression, in accordance with the Gibbs phase rule, the allotropic α/β-transition line $T_0 = T_0(x_A)$ for binary alloys can be deduced.

If, as is usual, decomposition occurs, the most favourable—that is the most negative—Gibbs energies are obtained by the common tangent at the

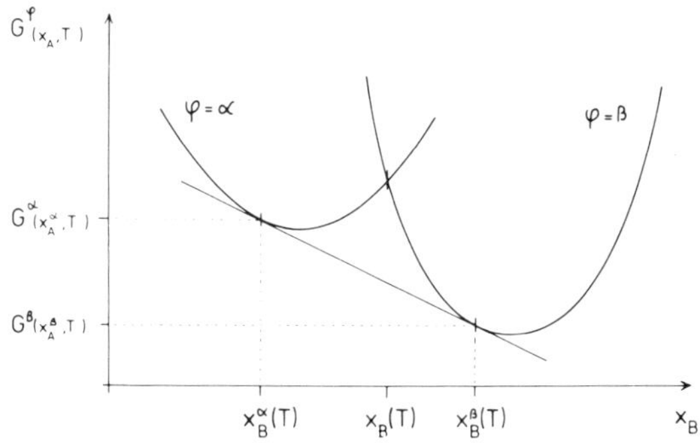

FIG. 8.   Gibbs energies of the phases α and β in relation to the alloy composition x_B at constant temperature (schematic).

two Gibbs energy curves G^α and G^β. This is illustrated in Fig. 8. The Gibbs energy of alloys with a composition x_B in the region x^α_B to x^β_B can diminish to a value G^{Gm}, where Gm corresponds to the common tangent, which is the sum of the two Gibbs energies of the phases α and β with the different compositions x^α_B and x^β_B. x^α_B and x^β_B are the limits of the two-phase region $(\alpha + \beta)$.

The compositions x^α_B and x^β_B can be determined by setting the partial Gibbs energies of each component equal to each other:

$$G^\alpha_A(x^\alpha_B, T) = G^\beta_A(x^\beta_B, T) \qquad \text{and} \qquad G^\alpha_B(x^\alpha_B, T) = G^\beta_B(x^\beta_B, T) \tag{17}$$

These partial Gibbs energies are defined as the final points of the tangent at the pure components. From Fig. 9 it follows directly that

$$G^\varphi_A = G^\varphi - x_B \frac{\partial G^\varphi}{\partial x_B} \qquad \text{and} \qquad G^\varphi_B = G^\varphi + (1 - x_B)\frac{\partial G^\varphi}{\partial x_B} \tag{18}$$

With the aid of the preceding relationships the equilibrium data of the

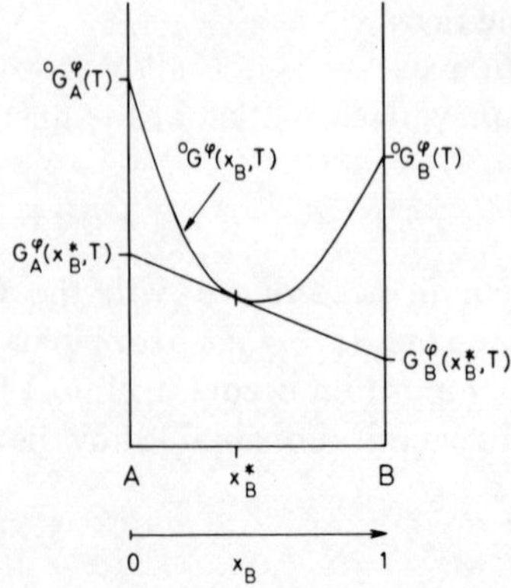

FIG. 9.   The partial Gibbs energies G_{A}^{φ} or G_{B}^{φ} of A or B, respectively, at composition x_{B}^{*} and temperature T (schematic).

phases α and β can be calculated, if the functions ${}^{E}G^{\alpha}$ and ${}^{E}G^{\beta}$ are known. Vice versa these relations give information about ${}^{E}G^{\alpha}$ and ${}^{E}G^{\beta}$ if the equilibrium data are known.

2.3 Chemical Potential and Chemical Activity

If a homogeneous phase φ is stable in a certain region of composition and temperature it is possible to deduce ${}^{E}G^{\varphi}$ directly from EMK—or partial pressure measurements. Such measurements first yield partial entities as the chemical potential, μ, or the chemical activity, a, which are related to the components. Therefore the connection between these partial entities and the integral Gibbs energies has to be formulated. From general rules of thermodynamics at constant pressure and temperature it follows that

$$G^{\varphi} = N_{A}\mu_{A}^{\varphi} + N_{B}\mu_{B}^{\varphi} \tag{19}$$

and thus the chemical potential of the A-atoms is defined by

$$\mu_{A}^{\varphi} = \left.\frac{\partial G^{\varphi}}{\partial N_{A}}\right|_{T,N_{B}\,=\,\text{const.}} \tag{20}$$

N_{A} or N_{B} are the numbers of A- or B-atoms in the alloy, respectively, with $N_{A} + N_{B} = N$.

The chemical potential of B is defined in a similar way. For pure A it follows from eqn (19), with $N_{A} = N$ and $N_{B} = 0$

$$ {}^{o}G_{A}^{\varphi}(T) = N\,{}^{0}\mu_{A}^{\varphi}(T) \tag{21}$$

A similar relation holds for B. By equalising eqn (15) with eqn (19) and using eqn (21), one obtains

$$^{E}G^{\varphi} = \sum_{K=A,B} N_K \{[\mu_K^{\varphi} - {}^{0}\mu_K^{\varphi}] - kT \ln x_K\} \tag{22}$$

(k is the Boltzmann constant). By this relation the chemical potentials of the alloy relative to those of the pure components in the same state of phase φ are connected with excess Gibbs energy. Relation (22) is further simplified if the chemical activities, a, of the components are introduced. These entities are defined by

$$kT \ln a_K^{\varphi} = \mu_K^{\varphi} - {}^{0}\mu_K^{\varphi} \qquad \text{with } K = A, B \tag{23}$$

One then obtains for $^{E}G^{\varphi}$

$$^{E}G^{\varphi} = kT \sum_{K=A,B} N_K \ln \frac{a_K^{\varphi}}{x_K} \tag{24}$$

The ratio of the chemical activity a_K^{φ} to the composition of the alloy x_K is called the activity coefficient of component K and phase φ. From the definition of the chemical activity in eqn (23) it follows for the reference state, here for the pure component, that

$$^{0}a_K^{\varphi}(T) = 1$$

One difficulty which must be mentioned, is when the chemical potentials or the activities of a certain phase cannot be measured for all compositions, for example, if the pure A-component exists in the ψ state instead of φ. In this instance only the difference

$$\mu_A^{\varphi}(x_B, T) - {}^{0}\mu_A^{\psi}(T)$$

is directly measurable. To obtain information about $^{E}G^{\varphi}$ with the aid of eqn (22) the following expression has to be used:

$$\mu_A^{\varphi}(x_B, T) - {}^{0}\mu_A^{\psi}(T) = \mu_A^{\varphi}(x_B, T) - {}^{0}\mu_A^{\varphi}(T) + \Delta^{0}\mu_A^{\psi/\varphi}(T)$$

It is seen that the left-hand side of the equation does not give any information about $^{E}G^{\varphi}$, if the difference $\Delta^{0}\mu_A^{\psi/\varphi}(T)$ is unknown. However, as has been mentioned before, these differences can be calculated for many different metals and different phases, (see, e.g. Fig. 7 and reference 10) using the relation

$$\Delta^{0}G_A^{\psi/\varphi}(T) = N\Delta^{0}\mu_A^{\psi/\varphi}(T)$$

14   W. PITSCH

where N is Avogadro's number and is equal to $6 \cdot 10^{23}\,\mathrm{mol}^{-1}$. With the aid of these data, $^{E}G^{\varphi}$, and therefore also G^{φ}, can be determined provided the alloy and the pure component are not in the same state of phase.

2.4 Illustration using an Atomistic Model

For binary AB alloys a simple expression for $^{E}G^{\varphi}$ can be obtained from the model of regular solutions. Using this model a random distribution of atoms on to the lattice sites of the φ-crystal is considered. It is assumed that the temperature dependent contributions to G^{φ} (the third term in eqn (5)) are already presented in eqn (15) by the contributions of the pure components. In addition, since a solid solution with random atom distribution is degenerated, there exists a zero point entropy (the second term in eqn (5)) which is equal to the third term in eqn (15). Then $^{E}G^{\varphi}$ follows solely from the zero point enthalpy of the solid solution (the first term in eqn (5)). This enthalpy is calculated with pairwise defined and temperature independent bond energies, V_{AA}, V_{BB}, V_{AB}, of nearest neighbour atom pairs. With the bond parameter $W^{\varphi} = 2V_{AB}^{\varphi} - V_{AA}^{\varphi} - V_{BB}^{\varphi}$ one obtains[11,12]

$$^{E}G^{\varphi} = \frac{NZ^{\varphi}}{2} x_{A} x_{B} W^{\varphi} \tag{25}$$

This excess term vanishes in the particular case where $2V_{AB}^{\varphi} = V_{AA}^{\varphi} + V_{BB}^{\varphi}$, i.e. $W^{\varphi} = 0$, which corresponds to the model of the ideal solution. However, in general, it is $2V_{AB}^{\varphi} \neq V_{AA}^{\varphi} + V_{BB}^{\varphi}$.

With the aid of eqns (15) and (25) it is possible to formulate the difference in the Gibbs energies of the solid solution, φ, and the mixture, Gm, of the pure components in the state φ:

$$\Delta G^{\mathrm{Gm}/\varphi}(x_{B}, T) = \frac{NZ^{\varphi}}{2}(1 - x_{B})x_{B}W^{\varphi} + NkT[(1 - x_{B})\ln(1 - x_{B}) + x_{B}\ln x_{B}] \tag{26}$$

There are two different cases to be considered: If $2V_{AB}^{\varphi}$ is more negative (or more positive) than $V_{AA}^{\varphi} + V_{BB}^{\varphi}$, that is if AB pairs are more strongly (or less strongly) bound than AA- and BB-pairs, then $W^{\varphi} < 0$ (or $W^{\varphi} > 0$). Correspondingly $\Delta G^{\mathrm{Gm}/\varphi}$ is composed of the negative (or positive) contribution of the enthalpy and the always negative contribution of the entropy (Fig. 10). In the first instance ($W^{\varphi} < 0$) it follows from eqn (26), that for all temperatures the $\Delta G^{\mathrm{Gm}/\varphi}$ curve has such a form, and that the solid solution is always more stable than the mixture of the pure components. In the second case ($W^{\varphi} > 0$) one obtains for different temperatures different curves which are shown in Fig. 11. It can be seen that for low temperatures

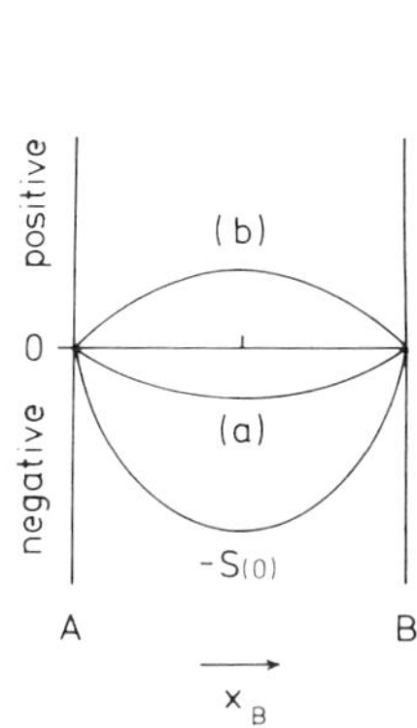

FIG. 10.   Enthalpies of eqn (26) for; (a) $W^\varphi < 0$ or (b) $W^\varphi > 0$, and entropy of eqn (26), $-S(0)$.

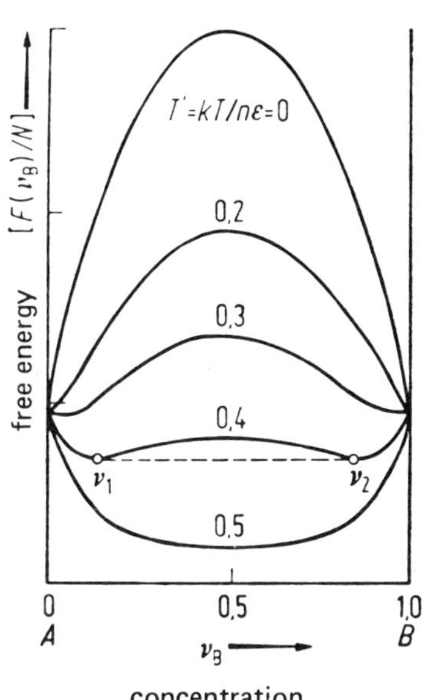

FIG. 11.   Free energies, which are equivalent to the Gibbs energies in eqn (26), for $W^\varphi > 0$ at different temperatures. The temperatures are indicated in a normalised form with $n\varepsilon = Z^\varphi W^\varphi/2$. The composition is indicated by v instead of x.[20]

the mixture of the pure components is more stable than the solid solution. However, with increasing temperature due to the influence of the entropy the solid solution becomes increasingly more stable. From the horizontal tangent, the solubility x_B^* can be deduced:

$$\left. \frac{\partial \Delta G^{\mathrm{Gm}/\varphi}}{\partial x_B} \right|_{x_B} = 0$$

This solubility line is presented in Fig. 12. For small solubilities, e.g. $x_B^* \ll 1$ on the A-rich side, one obtains approximately

$$x_B^*(T) = \exp\left(-\frac{Z^\varphi W^\varphi}{2kT} \right)$$

Above the critical temperature, $T_c = Z^\varphi W^\varphi/4K$, the increasing influence of the entropy means that the solid solution is at its most stable for the whole range of compositions.

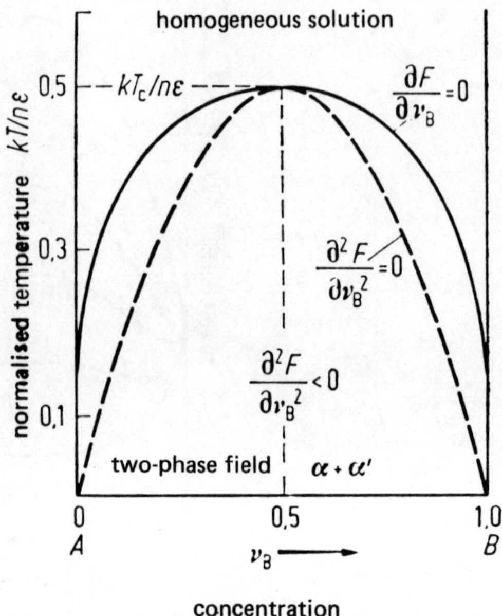

FIG. 12. Solubility gap (solid line) deduced from Fig. 11.[20] (The spinodal decomposition, indicated by the dashed line, is not treated in the text.)

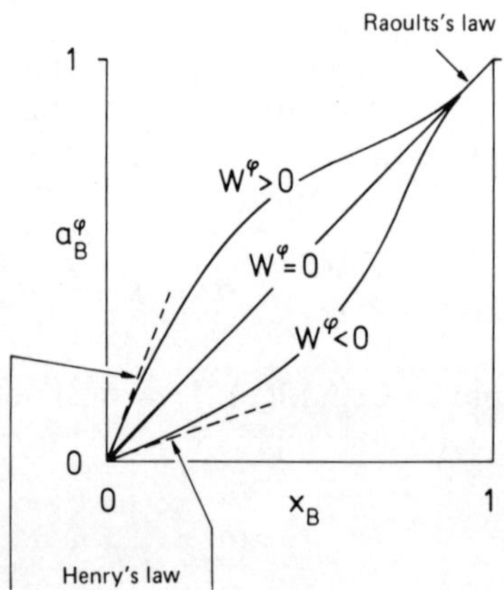

FIG. 13. Dependence of chemical activity, a, of the component B on alloy composition for an ideal solution ($W^\varphi = 0$) and solid solutions with ordering ($W^\varphi < 0$) or decomposing ($W^\varphi > 0$) tendencies.

To illustrate this the chemical activity of the component B can be formulated. Using eqns (23), (15), (18), (21) and (25) one obtains

$$kT \ln a_B^\varphi = kT \ln x_B + \frac{Z^\varphi W^\varphi}{4}(1 - x_B)^2$$

or

$$a_B^\varphi = x_B \exp\left[\frac{Z^\varphi W^\varphi}{4kT}(1 - x_B)^2\right]$$

This relationship is schematically drawn in Fig. 13. It can be seen, that under the conditions of ideal solutions, $W^\varphi = 0$ or $T \to \infty$, the chemical activity is equal to the alloy composition. If there is a tendency for ordering ($W^\varphi < 0$) the chemical activity is lowered. If there is a tendency for decomposition ($W^\varphi > 0$) the chemical activity is increased. The slope of the chemical activity curves at the pure components is given by

$$\left.\frac{da_B^\varphi}{dx_B}\right|_{x_B=0} = \exp\left[\frac{Z^\varphi W^\varphi}{2kT}\right]\begin{cases} <1 & \text{for } W^\varphi < 0 \\ >1 & \text{for } W^\varphi > 0 \end{cases}$$

or

$$\left.\frac{da_B^\varphi}{dx_B}\right|_{x_B=1} = 1 \qquad \text{for all } W^\varphi$$

If these slopes do not change within a certain range of alloy composition, it is said, that the component B follows the Henry or the Raoult law, respectively.

2.5 Application of the Model of Regular Solutions

After the equilibrium conditions between two solid solutions, $\varphi = \alpha, \gamma$, have been formulated with the parameters of the model of regular solutions, the limits of the two phase region x_B^α and x_B^γ can be deduced from eqns (15), (17), (18) and (25):

$$\Delta^0 G_A^{\alpha/\gamma}(T) + NkT \ln \frac{1 - x_B^\gamma}{1 - x_B^\alpha} = \frac{N}{2}[(x_B^\gamma)^2 Z^\gamma W^\gamma - (x_B^\alpha)^2 Z^\alpha W^\alpha]$$

and

$$\Delta^0 G_B^{\alpha/\gamma}(T) + NkT \ln \frac{x_B^\gamma}{x_B^\alpha} = \frac{N}{2}[(1 - x_B^\gamma)^2 Z^\gamma W^\gamma - (1 - x_B^\alpha)^2 Z^\alpha W^\alpha]$$

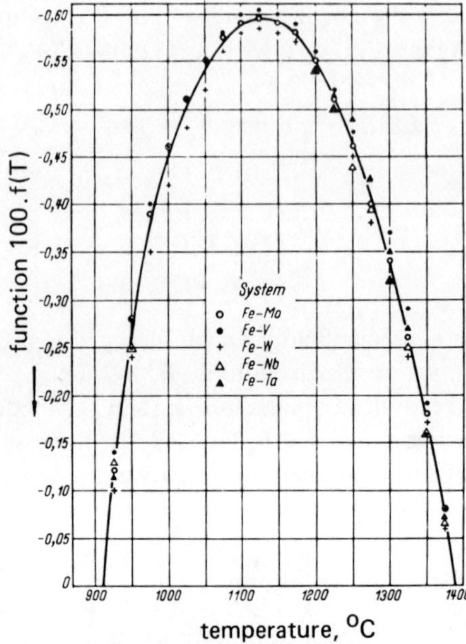

FIG. 14.   Width, $x_B^\alpha - x_B^\gamma$, of the $\alpha + \gamma$, two phase region (indicated as $100 \cdot f(T)$) as percentage atom fractions for different temperatures and different Fe–B systems.[14]

For the case where $x_B^\alpha \ll 1$ and $x_B^\gamma \ll 1$ a simpler expression is obtained:

$$x_B^\alpha - x_B^\gamma \approx \ln \frac{1 - x_B^\gamma}{1 - x_B^\alpha} \approx -\Delta^0 G_A^{\alpha/\gamma}(T)/NkT$$

From this relation it follows that the width, $x_B^\alpha - x_B^\gamma$, of the $\alpha + \gamma$ field is given solely by the difference of the Gibbs enthalpies of pure A in the α and γ states. This result, obtained earlier by W. Oelsen,[13] is well confirmed by measurements on iron alloys which have a closed γ-field and therefore only a small solubility for the second alloy component (see Fig. 14).[14]

It has to be emphasised that the model of regular solutions considers only solid solutions with random distributions of atoms. However for the case where $W^\varphi < 0$, short and long range ordered distributions of atoms are to be expected, at least at low temperatures. Due to this ordering effect the solid solution is more stable. In order to deal with this effect the model has to be extended further.[15]

2.6 General Methods

The model of regular solutions is very suitable for demonstrating the

TABLE 1
COMPILATION OF FREE ENERGY DIFFERENCE EQUATIONS
(Units, cal/g.at. and K)[a]

Iron

L–fcc	$3\,524 - 1{\cdot}959T$ above 1 665 K
L–bcc	$3\,300 - 1{\cdot}824T$ above 1 665 K
Bcc–fcc	$-1\,251{\cdot}2 + 2{\cdot}246\,8T - 0{\cdot}126\,55E - 2T^2 + 0{\cdot}220\,4E - 6T^3$ for $1\,100 \le T \le 1\,800$
Fcc–bcc	$1\,460 - 0{\cdot}827\,4T - 0{\cdot}178\,58E - 2T^2 + 0{\cdot}122\,5E - 5T^3$ for $300 \le T \le 1\,100$
Fcc–bcc	$1\,303 + 1{\cdot}78E - 3T^2 - 2{\cdot}87E - 5T^3 + 4{\cdot}91E - 8T^4$ below 300
Hcp–fcc	$-437 + 1{\cdot}12T$ above 300 K
Cobalt	
L–fcc	$3\,870 - 2{\cdot}19T$ above 1 500 K
Fcc–bcc	$-1\,662 + 0{\cdot}150\,9E - 2T^2 - 0{\cdot}670\,1E - 6T^3$ for $0 \le T \le 1\,800$
Fcc–bcc	-560 for $1\,300 \le T \le 1\,800$
Fcc–bcc	$-1\,840 + 1{\cdot}0T$ for $300 \le T \le 1\,300$

[a] $0{\cdot}126\,55E - 2 = 0{\cdot}126\,55 \times 10^{-2}$.

TABLE 2
COMPILATION OF EQUATIONS FOR EXCESS FREE ENERGY OF MIXING OF SOLUTION PHASES
(cal/g.at., K)[a]

Fe–Co

Liquid	$-1\,650 + 1{\cdot}10T$ for $1\,600 \le T \le 1\,900$
	$-3\,060 + 2{\cdot}96T$ for $1\,600 \le T \le 1\,900$
Fcc	$-555 + 0{\cdot}498\,3E - 3T^2 - 0{\cdot}102\,64E - 6T^3$ for $0 \le T \le 1\,800$
	$-235 + 0{\cdot}137\,03E - 2T^2 - 0{\cdot}282\,28E - 6T^3$ for $0 \le T \le 1\,800$
Bcc	$-8\,830 + 0{\cdot}119\,583E - 1T^2 - 0{\cdot}518\,664E - 5T^3$
	$-9\,533 + 0{\cdot}104\,454E - 1T^2 - 0{\cdot}395\,858E - 5T^3$
Hcp	$+345 + 0{\cdot}498\,3E - 3T^2 - 0{\cdot}102\,64E - 6T^3$ for $0 \le T \le 900$
	$+665 + 0{\cdot}137\,03E - 2T^2 - 0{\cdot}282\,28E - 6T^3$ for $0 \le T \le 900$

[a] The first line indicates $g(T)$, the second line $h(T)$ for each phase.

influence of the atomic bonding on the thermodynamic functions. Its application to the thermodynamics of solid solutions however, is limited, if additional effects such as chemical or magnetic ordering occur, as happens in iron and its alloys at lower temperatures. However, it has been found, that the excess Gibbs energy, given in eqn (25), can be extended in a purely formal way and then applied to many alloys:[5,16,17]

$$^{E}G^{\varphi} = x_{A}x_{B}[x_{A}g(T) + x_{B}h(T)]$$

In this expression temperature dependent functions $g(T)$ and $h(T)$ are parameters which have to be determined by application to known equilibrium data and/or known thermodynamic data.

As an example Tables 1 and 2 present the thermodynamic data of Fe–Co alloys for the α(bcc), γ(fcc) and ε(hcp) phases.[17] Energies which have been calculated from these data are presented for a temperature of 1000 K in Fig. 15. In this presentation for the sake of simplicity the Gibbs energies $^0G_{Fe}^\alpha$

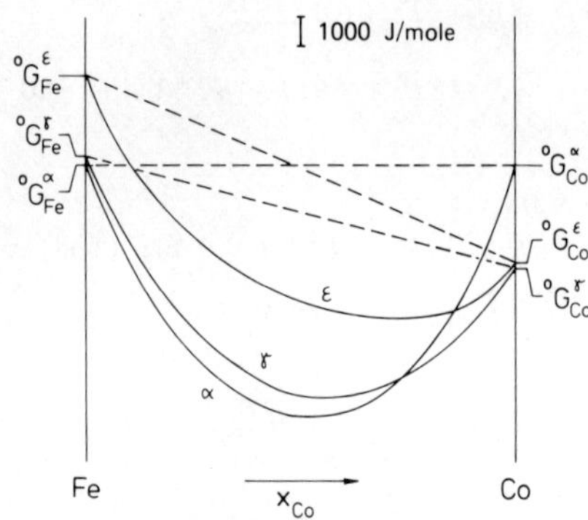

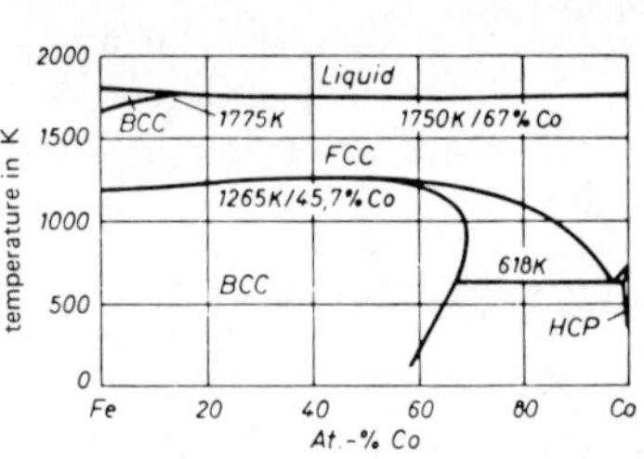

Fig. 15.   Gibbs energies of the α (bcc), γ (fcc) and ε (hcp) phases in Fe–Co alloys at 1000 K.

Fig. 16.   Calculated Fe–Co phase diagram.[17]

and $^0G_{Co}^\alpha$ were set equal. It is seen in the figure that the α-phase is most stable on the iron-rich side, and the γ-phase most stable on the cobalt-rich side The ε-phase is metastable for all alloy compositions. Similarly, it is possible, to calculate using the data in Tables 1 and 2, taken from reference 17, the whole Fe–Co diagram. The result of these calculations is shown in Fig. 16. Many binary and ternary alloys have been treated in a similar way.[10,21]

3.   INTERSTITIAL SOLID SOLUTIONS

Interstitial solid solutions are different from substitutional solid solutions in that the dissolved atoms are not exchangeable with the solvent atoms. Several thermodynamic models have been published in the literature for this case. They have been critically summarised in reference 18. The treatment which is most like the thermodynamic model of substitutional solid solutions, will now be discussed.[9]

In this treatment the interstitial sites are treated in a similar way to the lattice sites in substitutional solid solutions. It is convenient, to characterise the composition of the alloy by the ratio, y, of the number of interstitial atoms (e.g. C) to the number of interstitial sites. The latter is given by the number of solvent atoms (e.g. Fe) multiplied by a lattice dependent factor b^φ. Therefore

$$y_C = \frac{N_C}{b^\varphi N_{Fe}} = \frac{z_C}{b^\varphi} \quad \text{with } z_C = \frac{N_C}{N_{Fe}}$$

Hence the compositions of solid solutions lie within $0 < z_C < b^\varphi$, which means that the position of the carbon-rich reference state depends on the lattice structure φ.

The Gibbs energy of the interstitial solid solution, e.g. Fe–C, is then formulated as in eqn (15):

$$G^\varphi(z_C, T) = {}^0G^\varphi_{Fe}(T) + \frac{z_C}{b^\varphi}[{}^0G^\varphi_{FeC_b}(T) - {}^0G^\varphi_{Fe}(T)]$$

$$+ RTb^\varphi\left[\frac{z_C}{b^\varphi}\ln\frac{z_C}{b^\varphi} + \left(1 - \frac{z_C}{b^\varphi}\right)\ln\left(1 - \frac{z_C}{b^\varphi}\right)\right] + {}^EG^\varphi(z_C, T) \quad (27)$$

In this equation G^φ is related to 1 mol of solvent atoms. ${}^0G^\varphi_{FeC_b}$ is the Gibbs energy of 1 mol of fictitious groups of atoms FeC_b, the iron atoms of which build up the φ-crystal lattice. This atom group is the reference state on the carbon-rich side of the alloy. The factor b^φ is equal to the number of interstitial sides per solvent atom. The entropy of mixing corresponds in this case to the random distribution of the dissolved carbon-atoms on all interstitial sides. The excess Gibbs energy in this relation has been conveniently formulated in a similar way as for substitutional solid solutions as

$$^EG^\varphi(z_C, T) = \frac{z_C}{b^\varphi}\left(1 - \frac{z_C}{b^\varphi}\right)L^\varphi(T) \tag{28}$$

The temperature dependent functions ${}^0G^\varphi_{FeC_b}$ and L^φ are at first unknown. They have to be evaluated by adjustments to known data of thermodynamic quantities or phase equilibria. The conditions for these phase equilibria are obtained by equalising the partial Gibbs energies of the two components Fe and C, which are deduced from the Gibbs energy by means of the relationships

$$G^\varphi_{Fe} = G^\varphi - z_C\frac{\partial G^\varphi}{\partial z_C} \quad \text{and} \quad G^\varphi_C = \frac{\partial G^\varphi}{\partial z_C} \tag{29}$$

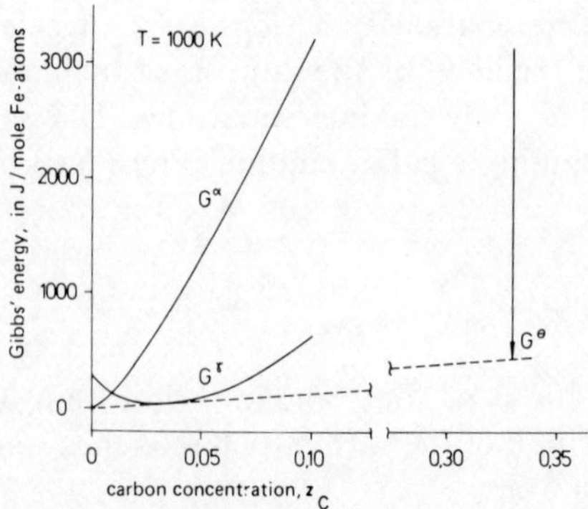

FIG. 17. Gibbs energies of the α (bcc), γ (fcc) and cementite ($\theta = FeC_{1/3}$) phases in the Fe–C system at 1000 K.

Then it follows that

$$G_{Fe}^{\varphi} = {}^0G_{Fe}^{\varphi}(T) + b^{\varphi}N_{Fe}kT\ln\left(1 - \frac{z_C}{b^{\varphi}}\right) + \left(\frac{z_C}{b^{\varphi}}\right)^2 L^{\varphi}(T) \qquad (30)$$

and

$$G_C^{\varphi} = {}^0G_C^{\varphi}(T) + N_{Fe}kT\ln\frac{z_C}{b^{\varphi} - z_C} - \frac{2z_C}{(b^{\varphi})^2}L^{\varphi}(T) \qquad (31)$$

with the abbreviation

$${}^0G_C^{\varphi}(T) = (b^{\varphi})^{-1}[{}^0G_{FeC_b}^{\varphi}(T) - {}^0G_{Fe}^{\varphi}(T) + L^{\varphi}(T)]$$

Since G^{φ} is related to 1 mol of solvent atoms, one can insert $N_{Fe}k = R$, (where R is the gas constant) into the preceding equations. By applying

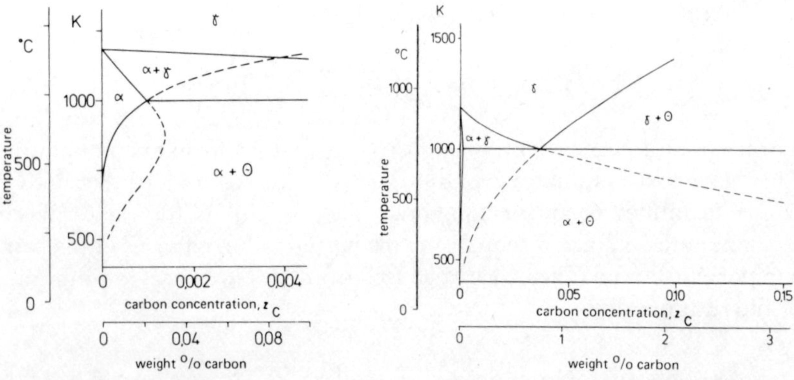

FIG. 18. Equilibrium lines of the α, γ and θ phases in Fe–C alloys.[19]

eqns (30) and (31) to the ferrite (α)/austenite (γ) equilibria of the Fe–C system, the Gibbs energies of these phases have been deduced.[19] An example is given in Fig. 17 which corresponds to 1000 K, the temperature of the perlite reaction. In Fig. 17 the ferrite–austenite two phase region is limited by $z_C^\alpha = 0{\cdot}000\,93$ and $z_C^\gamma = 0{\cdot}036$ (this corresponds to 0·02 Gew. % and 0·77 Gew. %). Similar calculations have been made in reference 19 for the whole range of temperatures between 800° and 1500 °C. The results obtained are presented in Fig. 18. The carbon solubility lines above 1000 K correspond to the measurements; below 1000 K they are extrapolated data. However the extrapolated lines are not simple geometrical curves, but follow from the thermodynamics of the Fe–C system.

4.   STOICHIOMETRIC COMPOUNDS

Compounds generally occur at certain stoichiometric compositions of the alloy. Their range of solubility is frequently very small. Therefore for many purposes it is sufficient, to consider these compounds solely at their stoichiometric compositions. Then the Gibbs energy is, as for the pure components, only a function of temperature (at constant pressure). Nevertheless, the quantitative evaluation of the Gibbs energy is only scarcely possible on the basis of eqn (5), since the necessary data are mostly not available. Therefore in general the Gibbs energies of these stoichiometric compounds are deduced from the equilibrium data for other phases. This will be shown, as an example, for cementite (θ) in the Fe–C system.

In Fig. 19 the Gibbs energies of cementite and of ferrite or austenite are sketched. z_C^φ is equal to the equilibrium composition of the solid solution, φ,

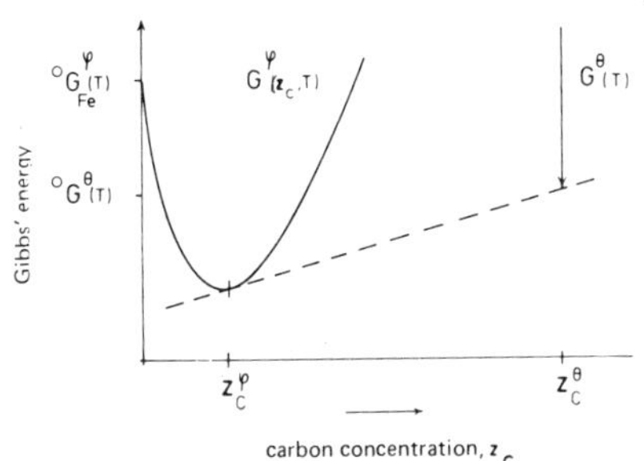

FIG. 19.   Gibbs energies of a solid solution, φ, and a stoichiometric compound, θ, versus alloy composition at constant temperature (schematic).

whereas cementite is accounted for at the constant composition $z_C^\theta = 1/3$. From Fig. 19 one immediately obtains the equilibrium relation

$$^0G^\theta(T) - \frac{1}{3}\frac{\partial G^\varphi}{\partial z_C}\bigg|_{z_C^\varphi} = G^\varphi(z_C, T) - z_C^\varphi\frac{\partial G^\varphi}{\partial z_C}\bigg|_{z_C^\varphi}$$

and then from eqn (29)

$$^0G^\theta(T) = G^\varphi_{Fe}(z_C^\varphi, T) + \tfrac{1}{3}G^\varphi_C(z_C^\varphi, T) \tag{32}$$

It is seen from this equation that the symbol θ is equivalent to the cementite formula $FeC_{1/3}$. Therefore

$$^0G^\theta(T) = {}^0G_{FeC_{1/3}}(T)$$

It follows from eqn (32) that the Gibbs energy of cementite can be evaluated from the knowledge of the thermodynamic φ-functions and the measured equilibrium data (z_C^φ, T). One obtains, for example, for the temperature of the perlite reaction, $^0G^\theta(1000\,K) = 630\,J/mol$ Fe atoms. This value is also indicated in Fig. 17 at the point $z_C^\theta = 0.33$. The common tangent with G^α and G^γ indicates the three phase equilibrium of the perlite reaction. On the other hand, if $^0G^\theta(T)$ is known for all temperatures it is possible to calculate the stable and metastable equilibria of ferrite and austenite with cementite. Inserting the expressions of eqns (30) and (31) into eqn (32) one obtains, with the approximation $z_C^\varphi \ll 1$

$$\left(\frac{z_C}{b^\varphi}\right)^{1/3} = \exp\left[\frac{^0G^\theta(T) - {}^0G^\varphi_{Fe}(T) - \tfrac{1}{3}{}^0G^\varphi_C(T)}{RT}\right]$$

The solubility lines obtained from this relation are also included in Fig. 18.

5.   SUMMARY

How the stability of pure metals, substitutional solid solutions, interstitial solid solutions and compounds can be analysed by means of thermodynamic functions, particularly by the Gibbs energies, has been discussed. The criterion for the mutual stability of two phases is deduced from the difference of the Gibbs energies of these phases. The phase with the more negative Gibbs energy is the more stable one. Knowing the quantitative data of the differences of the Gibbs energies provides the basis for calculating the forces, which promote a phase transformation and thereby a change in the structure. The literature referring to some of the most recent

TABLE 3
SUMMARY OF COMPUTER CALCULATIONS OF THE THERMODYNAMIC FUNCTIONS AND STATES OF EQUILIBRIA FOR Fe AND RELATED ALLOY SYSTEMS

Fe–C[a,m,v]	Fe–Nb[e]	Fe–Ni–Cr[t]	Cr–C[m]
Fe–N[b]	Fe–Mo[e,k]	Fe–Ni–Cu[t]	Mn–C[m]
Fe–Al[c]	Fe–Sn[l]	Fe–Mn–Cu[t]	Co–C[m]
Fe–Si[d]	Fe–W[e,k]	Fe–Mo–W[k]	Ni–C[m]
Fe–Ti[e]	Fe–Cr–C[o,v]	Fe–Cr–Mn[u]	Nb–C[m]
Fe–V[f]	Fe–Mn–C[p,v]	Co–Al[c]	Mo–C[m]
Fe–Cr[g]	Fe–Ni–C[q,v]	Co–Cr[g]	W–C[m]
Fe–Mn[h]	Fe–Mo–C[r,v]	Co–Ni[g]	Cr–N[n]
Fe–Co[e,g]	Fe–W–C[v]	Ni–Al[c]	Fe–Ni–Cr–Si[w]
Fe–Ni[g]	Fe–V–C[v]	Ni–Cr[g]	
Fe–Cu[i]	Fe–Cu–C[v]	Co–Si[d]	
Fe–Zn[j]	Fe–Co–Cr[s]	Ni–Si[d]	

[a] Harvig, H. (1971). *Jernkont. Ann.*, **155**, 157–61.

[b] Hillert, M. and Jarl, M. (1975). *Met. Trans.*, **6A**, 553–8.

[c] Kaufman, L. and Nesor, H. (1978). *Calphad*, **2**, 325.

[d] Kaufman, L. (1979). *Calphad*, **3**, 45.

[e] Kaufman, L. and Nesor, H. (1978). *Calphad*, **2**, 55.

[f] Hack, K., Nüssler, H. D., Spencer, P. J. and Inden, G. (1979). Berichte der Calphad VIII Konferenz, Stockholm, p. 244.

[g] Kaufman, L. and Nesor, H. (1973). *Z. Metallkd.*, **64**, 249.

[h] Kaufman, L. (1978). *Calphad*, **2**, 117.

[i] Harvig, H., Kirchner, G. and Hillert, M. (1972). *Met. Trans.*, **3**, 329.

[j] Kirchner, G., Harvig, H., Moquist, K. R. and Hillert, M. (1973). *Arch. Eisenhüttenwes*, **44**, 227.

[k] Kirchner, G., Harvig, H. and Uhrenius, B. (1973). *Met. Trans.*, **4**, 1059.

[l] Nüssler, H. D., von Goldbeck, O. and Spencer, P. J. (1979). *Calphad*, **3**, 19; and Ref. 6, p. 267.

[m] Kaufman, L. and Nesor, H. (1978). *Calphad*, **2**, 325.

[n] Jarl, M. (1977). *Calphad*, **1**, 201.

[o] Lundberg, R., Waldenström, M. and Uhrenius, B. (1977). *Calphad*, **1**, 159.

[p] Hillert, M. and Waldenström, M. (1977). *Met. Trans.*, **8A**, 5; and (1977). *Calphad*, **1**, 97.

[q] Rao, M. M., Russell, R. J. and Winchell, P. G. (1967). *Trans. AIME*, **239**, 634.

[r] Chatfield, C. and Hillert, M. (1977). *Calphad*, **1**, 159.

[s] Allibert, C., Bernard, C., Nüssler, H. D. and Spencer, P. J. Ref. 6, p. 207.

[t] Hasebe, M. and Mishizawa, T. (1977). 'Applications of Phase Diagrams in Metallurgy and Ceramics', Proceedings of NSB Workshop, Gaithersburg/USA, Vol. 2, App. 911–54.

[u] Kirchner, G. and Uhrenius, B. (1974). *Acta Met.*, **22**, 523.

[v] Uhrenius, B. (1978). In *Hardenability Concepts with Application to Steel* (Ed. D. V. Doane and J. S. Kirkaldy), Publication of the Metallurgical Society of AIME, pp. 28–81.

[w] Chart, T., Putland, F. and Dinsdale, A. Ref. 6, p. 183.

evaluations of the thermodynamics of iron and related alloy systems is summarised in Table 3. A more complete review of the thermodynamic calculations of phase diagrams, which also includes non-ferrous alloys, is found in reference 21.

REFERENCES

1. Zener, C. (1967). In *Phase Stability in Metals and Alloys*, (Ed. P. S. Rudman, J. Stringer and R. I. Jaffee) McGraw-Hill, New York, p. 25.
2. Hultgren, R. *et al.* (1973). *Selected Values of the Thermodynamic Properties of the Elements*, ASM.
3. Kaufman, L. (1959). *Acta Met.*, **7**, 575.
4. Weiss, R. J. and Tauer, K. J. (1956). *Phys. Rev.*, **102**, 1490.
5. Kaufman, L. and Bernstein, H. (1970). *Computer Calculations of Phase Diagrams*, Academic Press, New York/London.
6. Kaufman, L. (1971). *Metallurgical Chemistry*, 387.
7. Christian, J. W. (1965). *The Theory of Transformations in Metals and Alloys*, Pergamon Press, Oxford, p. 415.
8. Hillert, M. (1970). In *Phase Transformations*, ASM, pp. 181–218.
9. Hillert, M. 'Prediction of Iron-Base Phase Diagram' TRITA-MAC-0132, Royal Institute of Technology, Stockholm, December 1977; and Hillert, M. (1978). In *Hardenability Concepts with Applications to Steel*, (Eds. D. V. Doane and J. S. Kirkaldy) Publication of the Metallurgical Society of AIME, pp. 5–27.
10. Kaufman, L. and Bernstein, H. (1970). *Computer Calculation of Phase Diagrams*, Academic Press, New York/London.
11. Becker, R. (1955). *Theorie der Wärme*, Springer-Verlag, Berlin–Göttingen–Heidelberg, Sections 66 to 69.
12. Becker, R. (1937). *Z.f. Metallkde.*, **29**, 245–9.
13. Oelsen, W. and Wever, F. (1948). *Arch. Eisenhüttenwesen*, **19**, 97.
14. Fischer, W. A., Lorenz, K., Fabritius, H. and Schlegel, D. (1970). *Arch. Eisenhüttenwesen*, **41**, 489.
15. Inden, G. (1976). Proceedings of the Project Meeting Calphad V, Düsseldorf, FRG, p. IV 1.1–35.
16. Kaufman, L. (1972). *International Symposium on Metallurgical Chemistry* (Ed. O. Kubaschewski), HMSO, London, p. 373.
17. Kaufman, L. and Nesor, H. (1973). *Z. Metallkunde*, **64**, 249.
18. Ban-ya, S., Elliot, J. F. and Chipman, J. (1969). *Trans. AIME*, **245**, 1199.
19. Harvig, H. (1971). *Jernkont. Ann.*, **155**, 157.
20. Haasen, P. (1974). *Physikalische Metallkunde*, Springer-Verlag, Berlin–Heidelberg–New York, p. 82–101.
21. Ansara, I. (1979). *International Metals Reviews*, **1**, 20–53.

2

Special Computational Techniques

G. Inden

*Max Planck Institut für Eisenforschung GmbH,
Düsseldorf, FRG*

1. INTRODUCTION

The usual (temperature versus composition) phase diagram is a graphical representation of the most stable states of a condensed solution, at a given temperature and at constant pressure. With these conditions these states correspond to a minimum of the Gibbs energy of the solution. Calculation of a phase diagram can therefore be reduced to a mathematical problem to find the Gibbs energy minimum. This can be done numerically if the Gibbs energies of all competing phases are known as a function of temperature and composition. In practice, however, only the thermodynamic functions of observable phases, can be determined from experiments, and these in turn only in the temperature and composition ranges where they are stable. Therefore it is important to choose a suitable mathematical expression for the thermodynamic functions based upon realistic physical models and allowing for interpolation of the experimental data as well as for extrapolation outside the experimentally accessible ranges.

There are essentially two approaches to developing mathematical expressions for the thermodynamic functions. One is a formal polynomical series expansion with a sufficient number of adjustable parameters to obtain an adequate interpolation of experimental data. With an increasing number of parameters more and more complex physical effects can be treated. However, with this method a connection with physical models is possible only in a limited number of cases. The second approach is to divide the various contributions to the thermodynamic functions according to their underlying physical effects (e.g. anharmonic vibrations, atomic and/or magnetic ordering). The individual effects can then be treated separately with the required degree of sophistication or approximation by means of physical models. The advantage of this latter method is that contributions can be neglected or added without alteration to the remaining contributions. In the following presentation both procedures will be outlined and illustrated with simple or typical examples.

It should be mentioned that, in addition to the fundamental methods presented here, there also exist empirical or semi-empirical rules for predicting the appearance of particular phases in special alloys, e.g. the prediction of σ-phase appearance in superalloys ranging under the abbreviation 'Phacomp'. These rules have a much too limited application to be of general interest. Therefore these rules have been deliberately excluded from this presentation.

The methods of calculating phase equilibria have been known for a long time and calculated phase diagrams had already been published as early as 1908 (see Historical Introduction in reference 1). The computer-based methods, however, have been promoted essentially by L. Kaufman in the last two decades. A résumé of his work is published in reference 1. These methods are now applied throughout the world and a periodic journal *CALPHAD* (CALculation of PHAse Diagrams) has been published since 1977. An annual CALPHAD meeting has also been held since 1973. Numerous results of binary and ternary phase diagram calculations are presented at these meetings illustrating the usefulness and capabilities of the presently available programs. It is outside the scope of this presentation to give an overall view of all results already obtained in the past; instead, the underlying physical concepts and some aspects of numerical techniques will be discussed.

2.　TECHNIQUES FOR CALCULATING EQUILIBRIA OF SOLUTION PHASES

2.1 Polynomical Expression of Gibbs Energy for Substitutional Solutions

The Gibbs energy of one mole of solution in the structural state φ is usually written in the form

$$G_m^\varphi = \sum_J x_J^\varphi \, {}^0G_J^\varphi + {}^F G_m^\varphi \tag{1}$$

The first term is a weighted sum (according to the mole fractions x_J^φ) of the Gibbs energies of the pure components J in the structural state φ. This term is a reference frame for all energetic contributions due to the formation of the solution which are included in $^F G_m$. $^F G_m$ takes the simplest form if a random solution of non-interacting components is formed (ideal solution). Then $^F G_m$ reduces to the entropy term due to mixing:

$$^F G_m^\varphi = RT \sum_J x_J^\varphi \ln x_J^\varphi \tag{2}$$

The ideal solution is hypothetical. However, in practice the configurational entropy contribution to $^F G_m$ tends to this limiting expression with increasing temperatures and becomes the predominant contribution at high temperatures. Therefore, many solid–liquid phase equilibria for systems with weakly interacting components could be described in this approximation.[1] For non-ideal solutions one usually writes

$$^F G_m^\varphi = RT \sum_J x_J^\varphi \ln x_J^\varphi + {}^E G_m^\varphi \tag{3}$$

where $^E G_m^\varphi$ is the excess Gibbs energy incorporating all contributions due to deviations of real solutions from ideality. This term generally contains enthalpy ($^E H_m^\varphi$) and entropy ($^E S_m^\varphi$) contributions which may become complicated functions of temperature and composition. These may either be delined formally with a sufficiently elaborate mathematical expression or described by model calculations.

Most phase diagram calculations are performed at present with a power series in x and T of the general form

$$^E G_m^\varphi = \sum \sum_{I \neq J} x_I^\varphi x_J^\varphi [{}^0 L_{IJ}^\varphi + {}^1 L_{IJ}^\varphi (x_I^\varphi - x_J^\varphi)^1 + \cdots + {}^n L_{IJ}^\varphi (x_I^\varphi - x_J^\varphi)^n + \cdots] \tag{4}$$

with

$$^n L_{IJ}^\varphi = {}^n A_{IJ}^\varphi + {}^n B_{IJ}^\varphi T + {}^n C_{IJ}^\varphi T^2 + \cdots \tag{5}$$

The coefficients $^n L_{IJ}$ are called interaction coefficients. If for all $n \geq 1, {}^n L_{IJ} \equiv 0$ and $^0 L_{IJ} = \text{constant}$, $^E G_m$ contains only an enthalpy contribution. Then eqn (4) takes the form

$$^E G_m^\varphi = \sum \sum_{I \neq J} x_I^\varphi x_J^\varphi L_{IJ}^\varphi \tag{6}$$

which holds for a hypothetical random solution of pairwise interacting components (regular solution). A large number of systems has been reasonably well described with this approximation.[1] If for all $n > 1$, $^nL_{IJ} \equiv 0$ and $^1L_{IJ} = \text{constant}$ then

$$^EG_m^\varphi = \sum\sum_{I \neq J} x_I^\varphi x_J^\varphi [^0L_{IJ}^\varphi + {}^1L_{IJ}^\varphi(x_I^\varphi - x_J^\varphi)] \tag{7}$$

This form is called the subregular solution approach.*

As long as series (5) is truncated after the linear term in T, the coefficients $^nA_{IJ}$ and $^nB_{IJ}$ can be directly determined from experimental excess enthalpy and excess entropy data. If higher orders of T are considered, the coefficients enter into both the enthalpy and the entropy, thereby complicating their numerical determination from experiments.

A method of deriving higher order temperature dependences from physical models will be described later for the cases of atomic and magnetic ordering.

2.2 Methods of Phase Boundary Calculation

In this section it is presumed that the Gibbs energies of all competing phases, $\varphi = 1, 2, \ldots P$, are known. Let the number of components be c and let n_I^φ be the mole numbers of component $I = A, B, \ldots$ in the structural state φ. Then the Gibbs energy of a phase φ is

$$G^\varphi = G^\varphi(n_A^\varphi, n_B^\varphi, \ldots)$$

If a solution is composed of P different phases, then the Gibbs energy of the solution is

$$G = \sum_{\varphi = 1}^P G^\varphi(n_A^\varphi, n_B^\varphi, \ldots) \tag{8}$$

The equilibrium state is then given by the set of n_I^φ values ($c \cdot P$ unknowns)

---

* The advantage of the polynomial form, eqn (7), is that it reduces immediately to the regular solution form setting $^1L_{IJ} \equiv 0$. This is not true for the alternative but equivalent expression

$$^EG_m^\varphi = \sum\sum_{I \neq J} x_I^\varphi x_J^\varphi [x_I^\varphi g_{IJ}^\varphi + x_J^\varphi h_{IJ}^\varphi]$$

This expression yields eqn (7), setting $g_{IJ}^\varphi = {}^0L_{IJ}^\varphi + {}^1L_{IJ}^\varphi$ and $h_{IJ} = {}^0L_{IJ} - {}^1L_{IJ}$. For the regular solution (i.e. with $^1L_{IJ} \equiv 0$) one obtains $L_{IJ}^\varphi = (x_I^\varphi + x_J^\varphi)^0L_{IJ}$ which is not constant except for binaries as generally one has $x_I^\varphi + x_J^\varphi \neq 1$.[2]

which minimises eqn (8). This minimisation has to be carried out with the following constraints:

$$\sum_{\varphi=1}^{P} n_I^{\varphi} = n_I = \text{mole number of } I \text{ atoms} \qquad (9)$$

Instead of the $c \cdot P$ parameters n_I^{φ} one often prefers to introduce other variables: P parameters n^{φ} giving the amount of phase φ in moles and $c \cdot P$ parameters $x_I^{\varphi} = n_I^{\varphi}/n^{\varphi}$. Then the Gibbs energy of a certain amount of solution is

$$G = \sum_{\varphi=1}^{P} n^{\varphi} G_{\mathrm{m}}^{\varphi}(x_{\mathrm{A}}^1, x_{\mathrm{A}}^2, \ldots x_{\mathrm{B}}^1, \ldots)$$

2.2.1 Minimum Search Using Derivatives

The minimisation of eqn (8) with the constraints of eqn (9) can be performed with Lagrange's multipliers λ_I:

$$\frac{\partial G}{\partial n_I^{\varphi}} + \sum_J \lambda_J \frac{\partial \left[\sum_{\varphi} n_J^{\varphi} - n_J \right]}{\partial n_I^{\varphi}} = \frac{\partial G}{\partial n_I^{\varphi}} + \sum_J \lambda_J \frac{\partial n_J^{\varphi}}{\partial n_I^{\varphi}}$$

$$= \frac{\partial G}{\partial n_I^{\varphi}} + \lambda_I = 0 \qquad (c \cdot P \text{ eqn}) \qquad (10)$$

The first term is the chemical potential of component I in φ which is usually written as $\bar{G}_I^{\varphi}$. Since the λ_I are the same for all $\varphi = 1, \ldots P$, the system of eqn (10) can be reduced to the following system.

$$\bar{G}_I^1 = \bar{G}_I^2 = \cdots = \bar{G}_I^P \qquad I = \mathrm{A, B}, \ldots \qquad (11)$$

The total number of equations (eqn (11)) is $c \cdot \sum_{v=1}^{P-1} (P - v)$ but only $c \cdot (P - 1)$ of them are independent since eqn (11) has been obtained by elimination of the c unknowns λ_I in eqn (10). In order to determine the equilibrium state a set of $c \cdot (P - 1)$ equations like eqn (11) has to be solved. This can be done by numerical techniques such as the Newton–Raphson method.[3] It is also possible to rewrite eqn (11) to transform the solution problem to a problem involving a search for a minimum and then to use rapidly convergent steepest descent techniques.[4,5]

These methods require evaluation of derivatives. This makes any change in Gibbs energy (e.g. taking into account previously neglected effects)

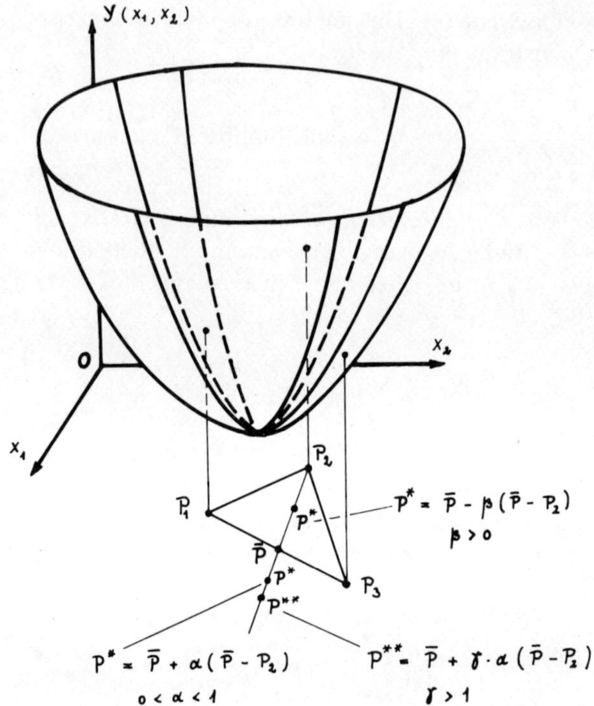

FIG. 1.　Illustration of the minimum search method by Nelder and Mead:[6] a two-dimensional example.

laborious. This complication is circumvented if so-called hill-climbing techniques are used.

2.2.2 Minimum Search Without Using Derivatives

This method developed by Nelder and Mead,[6] was first applied to Gibbs energy minimisation by Counsell *et al.*[7] The procedure will be briefly outlined for the general case and illustrated for a two-dimensional case (Fig. 1).

Consider a function $y(x_1, \ldots x_n)$ with a minimum. The procedure works as follows:

1. Define $n + 1$ points $P_1, \ldots P_{n+1}$ (P_1, P_2, P_3 in Fig. 1).
2. Determine P_h and P_1 with $y_h = \max(y_1, \ldots y_{n+1})$ and $y_1 = \min(y_1, \ldots y_1)$, ($P_h = P_2$ and $P_1 = P_3$ in Fig. 1).
3. Define centroid $\bar{P}$ of all P_i with $i \neq h$ ($\bar{P}$ is the middle of the line joining P_1 and P_3 in Fig. 1).

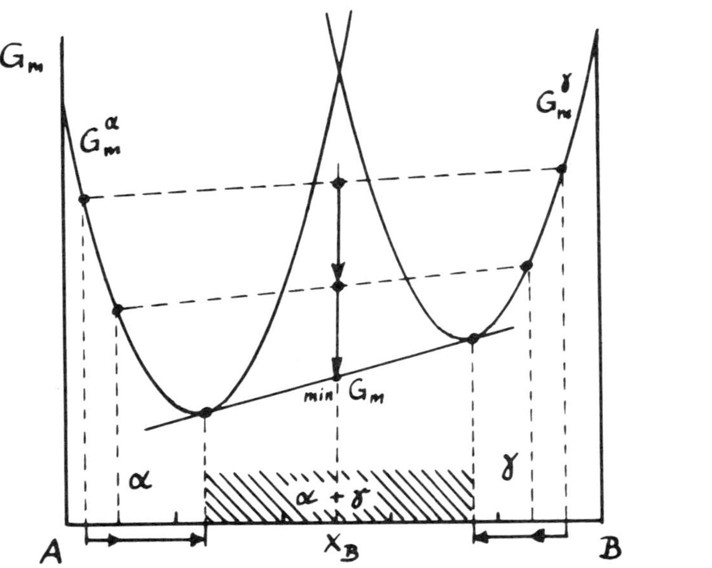

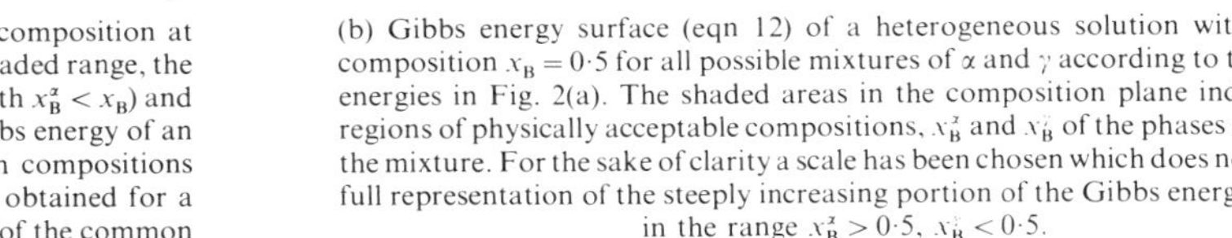

(a) Gibbs energies of the phase states α and γ, dependent on composition at constant temperature. For overall compositions, x_B within the shaded range, the Gibbs energy minimum is obtained with a mixture of phase α (with $x_B^\alpha < x_B$) and phase γ (with $x_B^\gamma > x_B$). The arrows indicate the variation of Gibbs energy of an equiatomic solution for mixtures of phase α and phase γ with compositions tending to the equilibrium values. The lowest Gibbs energy is obtained for a mixture with compositions corresponding to the contact points of the common tangent to the Gibbs energies of α and γ.

(b) Gibbs energy surface (eqn 12) of a heterogeneous solution with overall composition $x_B = 0.5$ for all possible mixtures of α and γ according to the Gibbs energies in Fig. 2(a). The shaded areas in the composition plane indicate the regions of physically acceptable compositions, x_B^α and x_B^γ of the phases α and γ in the mixture. For the sake of clarity a scale has been chosen which does not allow a full representation of the steeply increasing portion of the Gibbs energy surface in the range $x_B^\alpha > 0.5$, $x_B^\gamma < 0.5$.

Fɪɢ. 2.   Determination of the Gibbs energy minimum for the case of heterogenous equilibrium.

4. Set $P^* = \bar{P} + \alpha(\bar{P} - P_h)$ with $\alpha > 0$. P^* lies on the line $P_h\bar{P}$, on the far side of $\bar{P}$ from P_h.

5. If $y_1 < y^* < \max(y_i)$ for all $i \neq h$, P_h is replaced by P^* and the process restarted.

6. If $y^* < y_1$ an expansion is defined by $P^{**} = \bar{P} + \alpha\gamma(\bar{P} - P_h)$ with $\gamma > 1$.

7. If $y^{**} < y_1$, P^{**} replaces P_h and the process is restarted.

8. If $y^{**} > y_1$, P^* replaces P_h and the process is restarted.

9. If $y^* > \max(y_i)$ for all $i \neq h$, P^* yields a maximum as does P_h. Then P_h or P^* (depending on whichever yields the lowest y value) is taken to define a new $P^{**} = \bar{P} + \beta(\bar{P} - P_h)$ with $0 < \beta < 1$. P^{**} replaces P^* or P_h if $y^{**} < \min(y_h, y^*)$. In the failed case all P_i's are replaced by $(P_i + P_1)/2$ and the process is restarted.

10. The process stops when $\sqrt{\sum_i (y_i - \bar{y})^2/n} < \varepsilon$ with $\bar{y} = \sum_i y_i$.

The application of this method to the determination of the Gibbs energy minimum is illustrated in Fig. 2 for the case of a binary solution. Consider a mole of solution and two structural states α and γ. The Gibbs energies of the solution in the α or γ states are then G_m^α and G_m^γ. The most stable state of a solution with overall composition x_B can be obtained by considering the most general state of the solution to be composed of the amounts x^α and x^γ of the α and γ states. The corresponding Gibbs energy is then

$$G_m = x^\alpha G_m^\alpha + x^\gamma G_m^\gamma$$

$$= \frac{x_B - x_B^\gamma}{x_B^\alpha - x_B^\gamma} G_m^\alpha + \frac{x_B - x_B^\alpha}{x_B^\gamma - x_B^\alpha} G_m^\gamma \tag{12}$$

The minimisation of G_m is a two-dimensional problem with the variables x_B^α and x_B^γ. In Fig. 2(a) Gibbs energies for the structural states α and γ are schematically drawn for a case giving rise to a heterogeneous equilibrium. Figure 2(b) shows the G_m surface values (according to eqn (12)) for all possible points P_i, i.e. for all possible compositional combinations (x_B^α, x_B^γ). G_m exhibits a minimum at the point $(0\cdot25, 0\cdot75)$ as expected according to the energetic situation in Fig. 2(a).

3. SIMPLE CALCULATIONS OF PHASE DIAGRAMS

The mathematical formalism of phase diagram calculation having been described, it is now time to assemble the components necessary for performing a numerical calculation, i.e. to show how the numerical values of the entities in eqns (1) to (7) can be estimated or derived from experiments. This is best visualised starting with the simplest possible calculation of binary equilibria.

3.1 Ideal Solution Diagrams

Let us consider a solution with two competing structural states α and β (components A and B with mole fractions x_A^α, x_A^β, x_B^α, x_B^β). The Gibbs energy is then given by eqns (1) and (2) with $\varphi = \alpha, \beta$. From this the partial Gibbs energies (chemical potentials) can be directly obtained:

$$\begin{aligned} \bar{G}_A^\varphi &= {}^0G_A^\varphi + RT\ln x_A^\varphi \\ \bar{G}_B^\varphi &= {}^0G_B^\varphi + RT\ln x_B^\varphi \end{aligned} \qquad \varphi = \alpha, \beta$$

If G_m^α and G_m^β do not intersect, the equilibrium state is single phase given by the structural state of the lowest Gibbs energy. If, on the contrary, G_m^α and G_m^β do intersect, the most stable state is heterogeneous for the compositions between the tangency points of the common tangent to both functions (Fig. 2(a)). Therefore, the first information about the location of heterogeneous equilibria is given by the locus of intersection $x_0(T)$ or $T_0(x)$ defined by $G_m^\alpha = G_m^\beta$:

$$x_{0B}(T) = \frac{-\Delta^0 G_A^{\alpha\to\beta}}{\Delta^0 G_B^{\alpha\to\beta} - \Delta^0 G_A^{\alpha\to\beta}} \tag{13}$$

$$T_0(x_B) = \frac{x_A \cdot \Delta^0 H_A^{\alpha\to\beta} + x_B \cdot \Delta^0 H_B^{\alpha\to\beta}}{x_A \cdot \Delta^0 S_A^{\alpha\to\beta} + x_B \cdot \Delta^0 S_B^{\alpha\to\beta}} \tag{14}$$

The phase boundaries are obtained from the equilibrium conditions $\bar{G}_A^\alpha = \bar{G}_A^\beta$ and $\bar{G}_B^\alpha = \bar{G}_B^\beta$:

$$x_A^\beta = \frac{1 - \exp(\Delta^0 G_B^{\alpha\to\beta}/RT)}{\exp(\Delta^0 G_A^{\alpha\to\beta}/RT) - \exp(\Delta^0 G_B^{\alpha\to\beta}/RT)} \tag{15}$$

$$x_A^\alpha = x_A^\beta \cdot \exp(\Delta^0 G_A^{\alpha\to\beta}/RT) \tag{16}$$

(similarly for x_B^α, x_B^β).

Equations (13) to (16) indicate that the phase equilibria in ideal solutions are solely determined by the structural stabilities $\Delta^0 G_I^{\alpha\to\beta}$ of the components $I = $ A, B. Without knowing the whole set of these figures for all competing phases, not even the simplest phase diagram calculation can be started. It is therefore worth illustrating with one example how the presently used numerical values for the structural stabilities have been derived.

Kaufman[1] started with the simplest approach, assuming temperature independent enthalpies and entropies of transformation for pure metallic non-magnetic elements:

$$\Delta^0 G_I^{\alpha \to \beta} = \Delta^0 H_I^{\alpha \to \beta} - T . \Delta^0 S_I^{\alpha \to \beta} = a + b . T \tag{17}$$

This approach is applicable at temperatures above the Debye temperature θ_D since the predominant contribution in eqn (17) stems from differences in the phonon spectra of α and β which is nearly linear in T for $T > \theta_D$.[8] Differences in the electronic specific heat of α and β and in the anharmonic contributions to c_p ($c_p - c_v$ term) yield a T^2 term in eqn (17) which for typical cases is less than 10% of the linear term in the usual temperature ranges ($T < 3000$ K).

If α and β are stable phases, i.e. the transformation $\alpha \to \beta$ is experimentally observable, then in eqn (17) $\Delta^0 H^{\alpha \to \beta}$ is given by the heat of transformation, whereas $\Delta^0 S^{\alpha \to \beta} = \Delta^0 H^{\alpha \to \beta}/T_0^{\alpha \to \beta}$. In this way one could first start with the liquid–solid transformation if accurate data for the enthalpy of melting were available for all elements. Since these are still lacking for numerous (particularly high melting) elements Kaufman started his first assessment presuming Richard's rule[9] to be valid. This states that the entropy of melting of a stable close-packed structure (1st coordination 8–12) is $\simeq 8 \cdot 4$ J/mole/K. With this presumption, $\Delta^0 H^{\mathrm{sol} \to \mathrm{liq}}$ follows directly from the melting temperature.

The Pt–Re system will be considered as an example of this treatment. Figure 3(a) shows the phase diagram (thin, broken and solid lines) from Hansen[10] for which only the solid lines seem to be experimentally confirmed. The melting temperatures of the pure components are $T_{\mathrm{Pt}}^{\alpha \to \mathrm{L}} = 2042$ K and $T_{\mathrm{Re}}^{\varepsilon \to \mathrm{L}} = 3453$ K. With Richard's rule this immediately yields

$$\Delta^0 G_{\mathrm{Pt}}^{\alpha \to \mathrm{L}} = 8 \cdot 4(2042 - T)\,\mathrm{J/mole} \tag{18}$$

$$\Delta^0 G_{\mathrm{Re}}^{\varepsilon \to \mathrm{L}} = 8 \cdot 4(3453 - T)\,\mathrm{J/mole} \tag{19}$$

The unknown functions $\Delta^0 G_{\mathrm{Pt}}^{\varepsilon \to \mathrm{L}}$ and $^0 G_{\mathrm{Re}}^{\alpha \to \mathrm{L}}$ could similarly be derived if the hypothetical melting points $T_{\mathrm{Pt}}^{\varepsilon \to \mathrm{L}}$ and $T_{\mathrm{Re}}^{\alpha \to \mathrm{L}}$ were known. One can try to estimate them by a linear extrapolation from the regions where the transformations are observed. A first attempt is shown in Fig. 3(a), yielding $T_{\mathrm{Pt}}^{\varepsilon \to \mathrm{L}} = 1973$ K and $T_{\mathrm{Re}}^{\alpha \to \mathrm{L}} = 3373$ K. The proximity of the hypothetical melting points to the real ones implies that the entropy differences $\Delta^0 S_I^{\varepsilon \to \alpha}$ should be very small in order to avoid prediction of allotropic transformations $\varepsilon \rightleftarrows \alpha$ for Pt and Re which are not observed. Therefore setting $\Delta^0 S_I^{\alpha\to\varepsilon} = 0$ for both $I = $ Pt and Re, the Gibbs energies are then

$$\Delta^0 G_{Pt}^{\varepsilon\to L} = 8\cdot4(1973 - T)\,\text{J/mole}$$

$$\Delta^0 G_{Re}^{\alpha\to L} = 8\cdot4(3373 - T)\,\text{J/mole}$$

$$\Delta^0 G_{Pt}^{\varepsilon\to\alpha} = \Delta^0 G_{Pt}^{\varepsilon\to L} - \Delta^0 G_{Pt}^{L\to\alpha} = 8\cdot4(1973 - T - 2042 + T)$$
$$= -580\,\text{J/mole}$$

$$\Delta^0 G_{Re}^{\varepsilon\to\alpha} = 672\,\text{J/mole}$$

The last two terms are so small compared with RT that eqns (15) and (16) for the phase boundaries ε–α yield values greater than unity, i.e. physically non-acceptable solutions.

If, instead, lower melting temperatures are assumed as in Fig. 3(b), ($T_{Pt}^{\varepsilon\to L} = 1670\,\text{K}$ and $T_{Re}^{\alpha\to L} = 2890\,\text{K}$) one can try the following Gibbs energy functions:

$$\Delta^0 G_{Pt}^{\varepsilon\to L} = 9\cdot62(1670 - T)\,\text{J/mole} \tag{20}$$

$$\Delta^0 G_{Re}^{\alpha\to L} = 9\cdot62(2890 - T)\,\text{J/mole} \tag{21}$$

Equations (18)–(21) produce the $T_0^{\varepsilon\to L}$ and $T_0^{\alpha\to L}$ curves in Fig. 3(c). Combining these equations yields

$$\Delta^0 G_{Pt}^{\varepsilon\to\alpha} = -1080 - 1\cdot22T\,\text{J/mole} \tag{22}$$

$$\Delta^0 G_{Re}^{\varepsilon\to\alpha} = 1200 + 1\cdot22T\,\text{J/mole} \tag{23}$$

Introducing eqns (18)–(23) into eqns (15) and (16) gives the phase boundaries shown in Fig. 3(c). The final phase diagram is shown in Fig. 3(d), drawn in heavy lines, with the diagram from Hansen's book in thin lines.

There remain discrepancies between both diagrams which cannot be removed by choosing other stability parameters. The last set of $\Delta^0 G$ values is the one assessed by Kaufman[1] from an analysis of many Pt and Re diagrams. The remaining discrepancies should therefore be due to solution parameters which have arbitrarily been ignored in the ideal solution treatment. The above analysis demonstrates that the assessment of the Gibbs energy differences for the competing structural states has to be done by an iterative procedure which terminates when a consistent and satisfactory description of all available phase diagrams is obtained. The ideal solution treatment is not applicable to systems which exhibit large differences in the atomic size of the components or which form miscibility gaps or compounds. These have to be treated at least with a regular solution formalism.

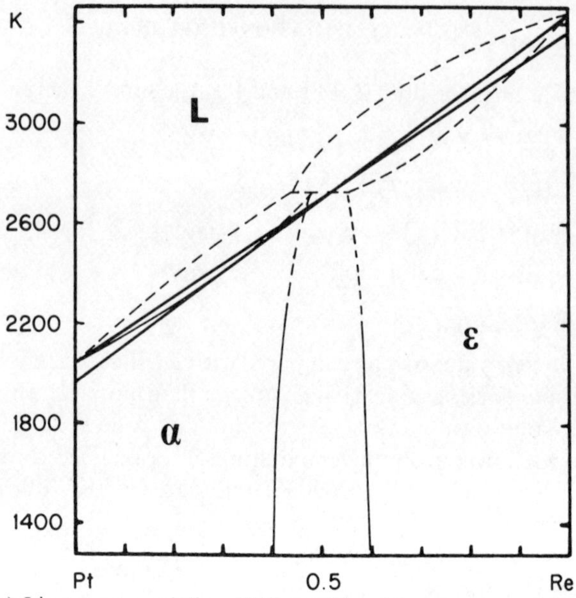

(a) Linear extrapolation of the melting temperatures of α-Pt (fcc structure) to pure Re and of ε-Re (hcp structure) to pure Pt, yielding the hypothetical melting points $T_{Re}^{\alpha \rightarrow L} = 3373\,\text{K}$ and $T_{Pt}^{\varepsilon \rightarrow L} = 1973\,\text{K}$.

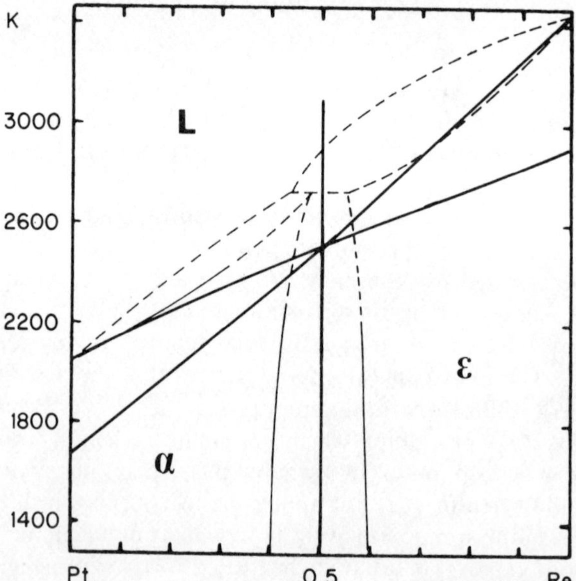

(b) Alternative linear extrapolation of the melting temperatures yielding $T_{Re}^{\alpha \rightarrow L} = 2890\,\text{K}$ and $T_{Pt}^{\varepsilon \rightarrow L} = 1670\,\text{K}$.

FIG. 3. Pt–Re phase diagram. Broken and thin lines according to Hansen's computation:[10] heavy lines according to the present calculation.

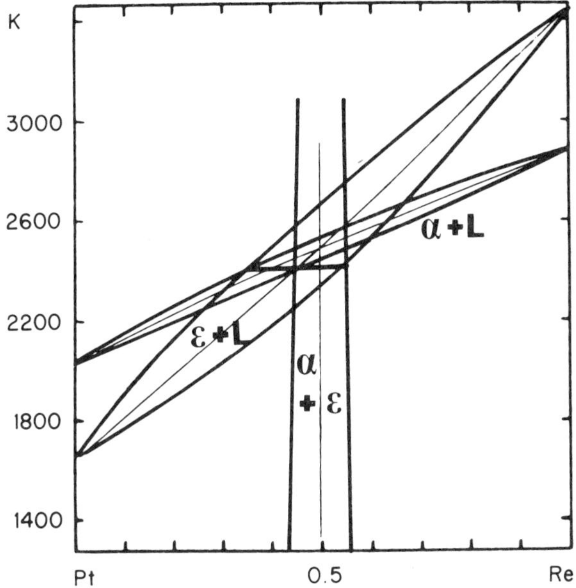

(c) Phase boundaries calculated with an ideal solution formalism using the extrapolated melting points in Fig. 3(b) for the equilibria α–L and ε–L.

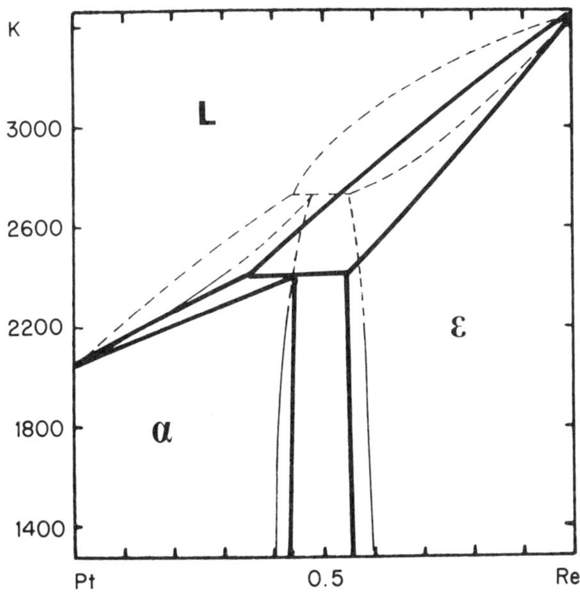

(d) Phase boundaries of the stable equilibria derived from Fig. 3(c) by omission of metastable equilibria.

FIG. 3—*contd.*

3.2 Regular Solution Diagrams

According to eqns (1), (3) and (6) the Gibbs energy of the regular solution is given by

$$G_m^\varphi = \sum_J (x_J^{\varphi 0} G_J^\varphi + RTx_J^\varphi \ln x_J^\varphi) + \sum \sum_{I \neq J} x_I^\varphi x_J^\varphi L_{IJ}^\varphi \tag{24}$$

From this one obtains for the partial Gibbs energies of a binary system

$$\bar{G}_A^\varphi = {}^0G_A^\varphi + RT\ln x_A^\varphi + L_{AB}^\varphi (x_B^\varphi)^2 \tag{25}$$

$$\bar{G}_B^\varphi = {}^0G_B^\varphi + RT\ln x_B^\varphi + L_{AB}^\varphi (x_A^\varphi)^2 \tag{26}$$

The $x_0(T)$ curve for a transition $\alpha \to \beta$ is obtained from the equation $G_m^\alpha = G_m^\beta$ which yields the quadratic relation

$$x_A \cdot \Delta^0 G_A^{\alpha \to \beta} + x_B \cdot \Delta^0 G_B^{\alpha \to \beta} + x_A \cdot x_B \cdot (L_{AB}^\beta - L_{AB}^\alpha) = 0$$

The equivalent $T_0(x)$ curve is given by

$$T_0(x_B) = \frac{(\Delta - x_B)\Delta^0 H_A^{\alpha \to \beta} + x_B \Delta^0 H_B^{\alpha \to \beta} + (1 - x_B)x_B(L_{AB}^\beta - L_{AB}^\alpha)}{(1 - x_B)\Delta^0 S_A^{\alpha \to \beta} + x_B \Delta^0 S_B^{\alpha \to \beta}}$$

This relation shows that according to the sign of $(L_{AB}^\beta - L_{AB}^\alpha)$ the T_0 curve is bent either upwards or downwards as compared to the ideal solution case. The phase boundaries are obtained by inserting eqn (25) and eqn (26) into the equilibrium conditions $\bar{G}_A^\alpha = \bar{G}_A^\beta$ and $\bar{G}_B^\alpha = \bar{G}_B^\beta$. This leads to a system of transcendental equations which have to be solved by the numerical techniques described in Section 2.2.

3.3 Determination of Regular Solution Parameters

The regular solution parameters of the stable phases can be obtained by experiment (e.g. enthalpies of mixing). However, the numerical values of all other competing phases are also needed. These have to be estimated from physical models or from empirical relations. These estimated values can be refined by variation until the best agreement with experimental information is obtained.

3.3.1 Kaufman's Estimation

Kaufman[11] split the interaction parameter of the liquid state into two terms:

$$L_{IJ}^{liq} = \varepsilon_0 + \varepsilon_p \tag{27}$$

The second term, the so-called internal pressure term, is taken from

Hildebrand-Scotts rule[12] which takes into account the size difference effects of the components:

$$\varepsilon_p = 0{\cdot}5(V_I + V_J)[(\Delta^0 H_I^{\text{liq}\to\text{vap}}/V_I)^{1/2} + (\Delta^0 H_J^{\text{liq}\to\text{vap}}/V_J)^{1/2}]^2 \qquad (28)$$

Here V_I, V_J are the molar volumes of the components I and J, and $\Delta^0 H_I^{\text{liq}\to\text{vap}}$ is the enthalpy of vaporisation of the pure component I.

The latter entity is temperature dependent and should be known for the temperature at which the liquid is considered. In view of the lack of data, however, Kaufman[11] suggested replacing the enthalpies and volumes in eqn (28) by their room temperature values and correcting the resulting overestimation by a global reduction of 40%.

The term ε_0 in eqn (27) stems from the chemical bonding in the liquid state. If constant pairwise interactions H_{IJ}^{liq} are presumed, the enthalpy of formation $^F H_{IJ}^{\text{liq}}$ of a random liquid solution can be written

$$^F H_{IJ}^{\text{liq}} = \frac{ZN}{2}[x_I H_{II}^{\text{liq}} + x_J H_{JJ}^{\text{liq}} + x_I x_J(2H_{IJ}^{\text{liq}} - H_{II}^{\text{liq}} - H_{JJ}^{\text{liq}})] \qquad (29)$$

with Z is equal to the coordination number of the liquid state. Setting $(ZN/2)H_{II}^{\text{liq}} = \Delta^0 H_I^{\text{liq}\to\text{vap}}$, a comparison of eqn (29) with eqns (27) and (24), yields directly

$$\varepsilon_0 = 2\frac{ZN}{2} H_{IJ}^{\text{liq}} - \Delta^0 H_I^{\text{vap}\to\text{liq}} - \Delta^0 H_J^{\text{vap}\to\text{liq}}$$

Kaufman[11] suggested setting

$$\frac{ZN}{2} H_{IJ}^{\text{liq}} = \Delta^0 H_M^{\text{vap}\to\text{liq}}$$

with M being the element in the periodic table exhibiting the group number $m = (i+j)/2$ corresponding to the average of the group numbers i and j of the components I and J. ($\varepsilon_0 = 0$ for elements in neighbouring columns of the periodic table.) A similar treatment has been proposed for solid structures:[13]

$$L^\varphi = L^{\text{liq}} + \varepsilon_0^\varphi + \varepsilon_p^\varphi$$

ε_p^φ is a strain-energy term:[14]

$$\varepsilon_p^\varphi = -0{\cdot}5(\Delta^0 H_I^{\varphi\to\text{vap}} + \Delta^0 H_J^{\varphi\to\text{vap}})(V_I - V_J)^2(V_I + V_J)^{-2}$$

and ε_0^φ depends on the difference in bond strength between the solid and liquid states:

$$\varepsilon_0^\varphi = 2\Delta^0 H_M^{\text{liq}\to\varphi} - \Delta^0 H_I^{\text{liq}\to\varphi} - \Delta^0 H_J^{\text{liq}\to\varphi}$$

Here M again stands for an element of the mean group number between I and J.

Using this method the interaction coefficients of a solution phase can be estimated from the enthalpies of sublimation and the structural stabilities of the components.

3.3.2 Miedema's Estimation

This method describes the enthalpy of formation of a solution in terms of Wigner–Seitz cells of the component elements. The enthalpy of formation originates from the change in boundary conditions when dissimilar atoms are brought into contact.

The model accounts for the enthalpy of formation of liquid metals,[14] binary solid alloys and intermetallic compounds with one or two transition metals.[15] It has been shown that the description and the predictions of the enthalpies of formation are sufficiently accurate to be of practical use. Details of this treatment cannot be given here.

3.3.3 Regular Solution Parameters from Low Temperature Equilibria

In a model of pairwise interactions, which do not sensibly depend on temperature and composition, the regular solution parameter L_{IJ}^{φ} originates from the difference in bond strength between like and unlike atom pairs. Therefore one can directly express L_{IJ}^{φ} in terms of these differences in the following way:[16]

$$L_{IJ}^{\varphi} = -\frac{N_L}{2}\left(z^{(1)}W_{IJ}^{(1)} + z^{(2)}W_{IJ}^{(2)} + \cdots\right) \qquad (30)$$

where N_L is Avogadro's number, $z^{(i)}$ is the coordination number of the ith shell and $W_{IJ}^{(i)}$ is the interchange energy between the atoms I and J as ith neighbours:

$$W_{IJ}^{(i)} = -2V_{ij}^{(i)} + V_{II}^{(i)} + V_{JJ}^{(i)}$$

$V_{IJ}^{(i)}$ is the bond energy of an atom pair $I\text{–}J$.

At first sight eqn (30) increases only the number of parameters instead of the single one L_{IJ}^{φ}. In practice, however, the parameters $W_{IJ}^{(i)}$ can be determined separately if ordering reactions between ith neighbours are observed.[16] These are expected to occur generally if the parameters $W^{(i)}$ are positive or special combinations of positive and negative values occur. In many instances (e.g. Ni–Al, Co–Al, Fe–Ti, Fe–Co–Si), ordered structures are stable up to the melting point. In all these cases L_{IJ}^{φ} can only be determined via a determination of $W_{IJ}^{(i)}$ since the hypothetical random alloy cannot be achieved experimentally.

An analysis of numerous bcc and fcc alloys[16,17] has shown that consideration of $W_{IJ}^{(1)}$ and $W_{IJ}^{(2)}$ is necessary and in most cases sufficient.

4. EXAMPLE OF A PHASE DIAGRAM CALCULATION WITH HIGHER ORDER POLYNOMIALS: Fe–V*

The following experimental information is available: The Gibbs energies of formation for the liquid state at $T = 2193$ K ($^F G_{FeV}^{liq}(2193)$) are known from an evaluation of activity measurements by Geiger $et\ al.$[18] No enthalpy or entropy data for the liquid state are available. In the bcc solid state α (here α stands for the paramagnetic bcc random solid solution) enthalpies of formation $^F H_{FeV}^{\alpha}$ have been determined at $T = 1623$ K by Spencer and Putland.[19] At this stage the thermodynamic data both for the liquid and for the solid state are incomplete, even at a single temperature. In order to discuss the phase equilibria the temperature dependences must also be known. Consequently some assumptions have to be made. At first it is assumed that the enthalpies and entropies of formation are independent of temperature. This presumption cannot be valid in reality, due, for example, to expected ordering reactions (short range ordering at high temperatures as a consequence of observed long range order at low temperatures[20]). However, it becomes more and more valid as the temperature is increased.

Even with this presumption one is still left with the unknown quantities $^F H_{FeV}^{liq}$ or $^F S_{FeV}^{liq}$ and $^F G_{FeV}^{\alpha}$ or $^F S_{FeV}^{\alpha}$. These represent too many unknown quantities to be determined by means of the known α–liquid phase boundaries.[10] The number of unknown quantities can be reduced to one if it is presumed that $^F H_{FeV}^{\alpha} = {}^F H_{FeV}^{liq}$. Then $^F S_{FeV}^{liq}$ is also fixed and there remains only $^F S_{FeV}^{\alpha}$ as an unknown quantity which can then be determined using the α–liquid phase boundaries. This procedure is adopted here. From a physical point of view it is then presumed that the chemical short range order contribution to $^F H$ is the same in the liquid as in the solid state (within the limits of accuracy of the experimental values). Instead of the above assumption another one can be made, namely $^F S^{liq} \equiv 0$, which also reduces the unknown quantities to a single one: $^E S^{\alpha}$. This assumption is rejected in

---

* This analysis[21] has been performed in cooperation with K. Hack, D. Nüssler and P. J. Spencer from Lehrstuhl f. Theor. Hüttenkunde, TH-Aachen.

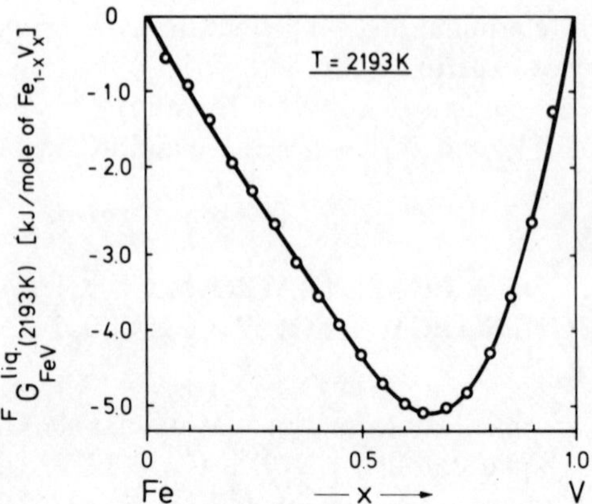

FIG. 4.   Gibbs energy of formation of liquid Fe–V alloys at $T = 2193$ K. The circles are the experiments of reference 18 and the solid line is a fifth order polynomial fitting curve (Appendix I).

the present instance since the whole treatment is much more sensitive to small changes in entropy (via the factor T) than in enthalpy.

At lower temperatures the σ phase becomes stable in the Fe–V system.[10] In a limited composition range around $x_v = 0.5$ the enthalpies of transformation $\Delta H^{\alpha \to \sigma}_{FeV}$ are known from experiment.[19] From the enthalpies, the entropies $\Delta S^{\alpha \to \sigma}_{FeV}$ can be estimated by means of the transformation temperature.[19] In the following, these data are used to develop $^F G^{\sigma}_{FeV}$ and $^F H^{\sigma}_{FeV}$ and to calculate the α–σ phase boundaries.

4.1 Analysis of the α–liquid Equilibria

4.1.1 Liquid State

The experimental Gibbs energies of formation at $T = 2193$ K are shown in Fig. 4 by open circles. The interpolation of these data is shown by a fifth order polynomial which is drawn as a solid line in Fig. 4 and tabulated in Appendix I.

The expression for $^F H^{liq}$ (which is set equal to $^F H^\alpha$) is obtained by fitting a third order polynomial to the experimental values of reference 19, Table I and Fig. 5. $^F S^{liq}$ is then immediately given by $^F H^{liq} - {}^F G^{liq}_{(2193)}/2193$. In Appendix I the resulting polynomial for $^E S^{liq}$ is given and this function is shown in Fig. 6. It is worth noting that the sign of $^E S^{liq}$ turns out as expected for short range order and magnetic contributions. The magnitude turns out to be rather small, smaller than the error band which is produced if the full error width of both $^F H^\alpha$ and $^F G^{liq}$ is used up. Therefore this result will not be

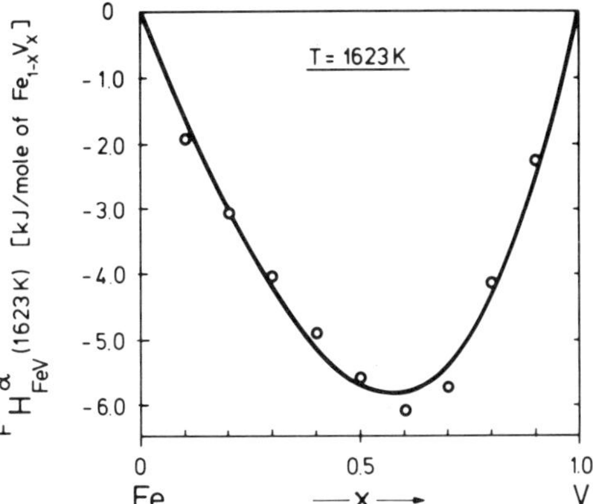

FIG. 5.   Enthalpy of formation of bcc (α) Fe–V alloys at $T = 1623$ K. The circles are the experiments of reference 19 and the solid line is the polynomial fitting curve of Appendix I. This enthalpy is considered to be temperature independent within the accuracy limits of the measurements. It is also used for the unknown enthalpy of formation of the liquid state.

discussed further; in particular no significance is attributed to the behaviour at $x_v > 0.75$.

4.1.2 Solid State

For the solid state, α, only $^F H^{\alpha}$ is taken from experiments and the polynomial has already been discussed above. The unknown entity $^E S^{\alpha}$ is

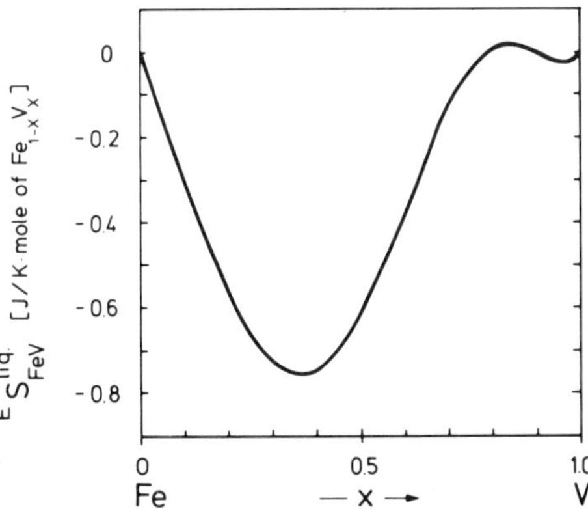

FIG. 6.   Entropy of formation of liquid Fe–V alloys as derived from the polynomials in Figs 4 and 5. This entropy is considered temperature independent in the range of 1600 K to 2200 K.

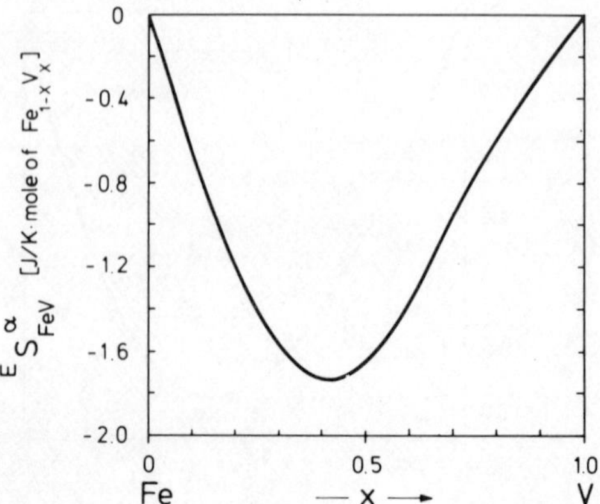

FIG. 7.	Entropy of formation of bcc (α) Fe–V alloys derived from the polynomial formulation of the Gibbs energy of the liquid state and from known α–liq. phase boundaries.

formulated as a fifth order polynomial, the coefficients of which are determined from the α–liq. phase boundaries. Here the effort was focussed mainly on obtaining the calculated azeotropic point as near to the experimental one as possible, thereby accepting some deviation from the experimental solidus–liquidus lines at other compositions (which are probably less accurate than the azeotropic point). It should be noticed that in this analysis not only the thermodynamic functions of the alloys but also those of the pure components have to be used in the calculations.

In the present analysis the functions $\Delta G_{Fe}^{\alpha \rightarrow liq}$ and $\Delta G_{V}^{\alpha \rightarrow liq}$ given by Kaufman[1] (Appendix II) have been used. The final result of the numerical treatment is given in Appendix I and Fig. 7. Again the computed result for $^{E}S^{\alpha}$ is negative as expected.

4.1.3 α–liquid Equilibrium

In Fig. 8 the Gibbs energies of the α and liquid states are shown for $T = 1775\,\mathrm{K}$, which is close to the azeotropic temperature. The phase boundaries resulting therefrom are shown in Fig. 9 together with the scarce experimental data which are satisfactorily reproduced by the calculations.

4.2 Analysis of the α–σ Equilibria

4.2.1 Thermodynamic Functions of the σ Phase

In a limited composition range around $x_v = 0.5$ the enthalpies of transformation $\Delta H_{FeV}^{\alpha \rightarrow \sigma}$ have been determined by Spencer and Putland.[19] At $x_v = 0.48$ the maximum transformation temperature occurs. At this

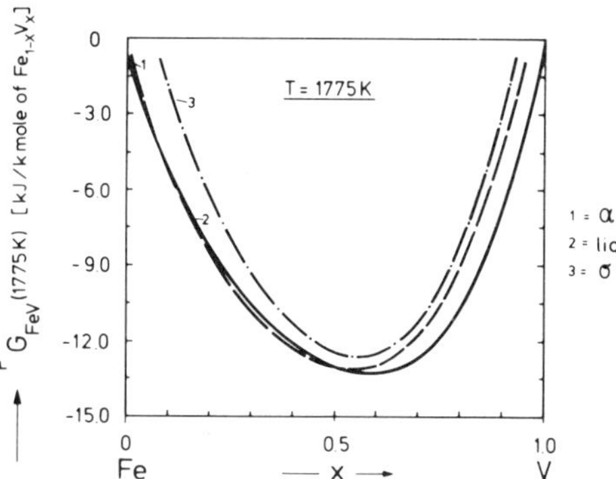

FIG. 8. Gibbs energies of formation for the bcc (α) state, the liquid state and for the σ phase at $T = 1775$ K.

composition the entropy of transformation is immediately obtainable from $\Delta S^{\alpha\to\sigma}_{(0\cdot48)} = \Delta H^{\alpha\to\sigma}_{(0\cdot48)}/T^{\alpha\to\sigma}_{(0\cdot48)}$. At compositions not too far away from $x_v = 0\cdot48$ the same evaluation holds approximately.

The enthalpies of formation of the σ phase can then be derived from the relation

$$^{F}H^{\sigma}_{FeV} = {}^{F}H^{\alpha}_{FeV} + \Delta H^{\alpha\to\sigma}_{FeV} - x_{Fe}\Delta^{0}H^{\alpha\to\sigma}_{Fe} - x_{V}\Delta^{0}H^{\alpha\to\sigma}_{V} \qquad (31)$$

The first term of the right-hand side is known.[19] The second term is known

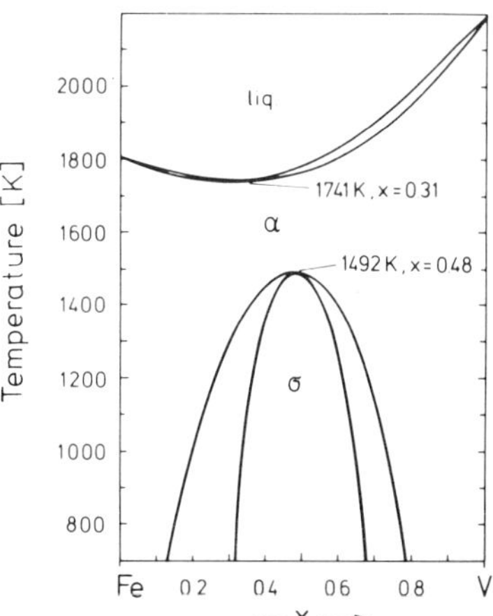

FIG. 9. Calculated α-liq. and α–σ phase boundaries of Fe–V alloys. The α-γ equilibria at the Fe-side have not been included in the calculations; they are therefore omitted here.

for some compositions.[19] The third term is taken from reference 22 (see Appendix II). The lattice stability of σ-Fe against α-Fe stems from an analysis of the α–σ equilibria in Fe–Cr. Here some doubts remain as to the α state: it is not clear whether the value applies to paramagnetic or magnetically short range ordered α-Fe. The value used here is in agreement with a value derived by Shao and Machlin.[23] These authors give $\Delta^0 H^{\alpha \to \sigma}_{Fe} = 3477 \, J/mol$, magnetic effects excluded. Considering the stabilisation of α-Fe by magnetic short range order (see Section 5) at $T^{\alpha \to \sigma}_{(0.48)} = 1490 \, K$, mag. $({}^0 H_{Fe}(1490 \, K)) -$ mag. $({}^0 H_{Fe}(T = \infty)) = -725 \, J/mol$, one gets $\Delta^0 H^{\alpha \to \sigma}_{Fe}(1490 \, K) \simeq 4200 \, J/mol$, which is very close to the value in Appendix II.

The last term in eqn (31) is completely unknown. In order to produce a figure for this value, the enthalpy of transformation $\Delta H^{\alpha \to \sigma}_{FeV}$ (which is known at $x_v = 0$ and at compositions around $x_v = 0.48$) is described by a second order polynomial (Appendix I) using the condition that $\Delta H^{\alpha \to \sigma}_{FeV}$ exhibits its minimum at $x_v = 0.48$. Then, for $\Delta^0 H^{\alpha \to \sigma}_v$, the value in Appendix II is obtained from the polynomial expression at $x_v = 1$.

A completely analogous treatment has been carried out for the entropy leading to the corresponding results for $\Delta S^{\alpha \to \sigma}_{FeV}$ (Appendix I) and $\Delta^0 S^{\alpha \to \sigma}_v$ (Appendix II).

With the thermodynamic expressions thus obtained the Gibbs energy for the σ phase is immediately derived (Appendix I).

5. ATOMIC AND MAGNETIC ORDERING

Any regular solution undergoes ordering or segregation reactions at sufficiently low temperatures, due to the difference in bond strength between like and unlike atom pairs. Similarly, magnetic ordering reactions occur in solutions with magnetic moment bearing atoms. Lowering T from high temperatures first produces short range ordering (sro) then at a critical temperature a periodic array of atoms or magnetic moments starts to appear called long range ordering (lro).

It is very difficult to incorporate physical models, which describe these reactions as a function of temperature, into the previously described formalism. This is due to the generally complicated expressions for the enthalpy and entropy in terms of long and short range order parameters. Therefore a complicated minimum search has to be done with respect to the order parameters (internal equilibrium) before the most stable state with respect to composition can be determined by the above numerical methods (external equilibrium).

It is therefore preferable to try an empirical approach.[24,25] It has been shown by Inden[25] that the contribution to the specific heat due to atomic or magnetic ordering reactions can be written as

$$c_p^{\text{lro}} = K^{\text{lro}} . R . \ln \frac{1 + \tau^3}{1 - \tau^3} \qquad \text{with } \tau = T/T_c < 1 \qquad (32)$$

$$c_p^{\text{sro}} = K^{\text{sro}} . R . \ln \frac{1 + \tau^{-5}}{1 - \tau^{-5}} \qquad \text{with } \tau > 1 \qquad (33)$$

R is Avogadro's number and T_c is the critical temperature for the long range ordering reaction involved. The constants K^{lro} and K^{sro} depend on T_c and on the total entropy difference, $\Delta S^{\text{ord} \to \text{dis}}$, between the disordered and fully ordered states in the following form

$$K^{\text{sro}} = \frac{0 \cdot 784 . f . \Delta S^{\text{ord} \to \text{dis}} / R}{0 \cdot 598 - 0 \cdot 211 . f}$$

$$K^{\text{lro}} = \frac{\Delta S^{\text{ord} \to \text{dis}} / R - 0 \cdot 493 . K^{\text{sro}}}{0 \cdot 822}$$

with

$$f = \int_{T_c}^{\infty} c_p^{\text{sro}} \, dT \Big/ \left(\int_0^{T_c} c_p^{\text{lro}} \, dT + \int_{T_c}^{\infty} c_p^{\text{sro}} \, dT \right)$$

i.e. the fraction of the total ordering energy due to sro. It has been found empirically[25] that for magnetic ordering $f = 0 \cdot 4$ (for bcc materials) and $0 \cdot 28$ (for fcc materials). In the case of atomic ordering $f \simeq 0 \cdot 2$ seems to be a good value for both fcc and bcc solid solutions.

5.1 Determination of the Enthalpy of Ordering

In order to obtain the enthalpy expressions (32) and (33) must be integrated. For the present purposes it is useful to separately evaluate the enthalpies:

$$\Delta H_{(T)}^{\text{lro}} = \int_0^T c_p^{\text{lro}} \, dT \qquad \text{for } T < T_c$$

$$\Delta H_{(T)}^{\text{sro}} = \int_{T_c}^T c_p^{\text{sro}} \, dT \qquad \text{for } T > T_c$$

The integrals yield[25] for $\tau = (T/T_\mathrm{c}) < 1$

$$\Delta H_{(T)}^{\mathrm{lro}} = K^{\mathrm{lro}} . R . T_\mathrm{c} \left\{ (1 - \tau) . \ln(1 - \tau) + \tau . \ln\left(\frac{1 + \tau^3}{1 + \tau + \tau^2} \right) \right.$$

$$+ \ln\left(\frac{1 + \tau}{\sqrt{\tau^4 + \tau^2 + 1}} \right) + \sqrt{3} . \arctan\left(\frac{2\tau - 1}{\sqrt{3}} \right)$$

$$\left. - \sqrt{3} . \arctan\left(\frac{2\tau + 1}{\sqrt{3}} \right) + 2 . \sqrt{3} . \arctan\left(\frac{1}{\sqrt{3}} \right) \right\} \qquad (34)$$

and for $\tau > 1$

$$\Delta H_{(T)}^{\mathrm{sro}} = K^{\mathrm{sro}} . R . T_\mathrm{c} \left\{ (\tau + 1)\ln(\tau + 1) - (\tau - 1)\ln(\tau - 1) \right.$$

$$+ \tau . \ln \frac{1 - \tau + \tau^2 - \tau^3 + \tau^4}{1 + \tau + \tau^2 + \tau^3 + \tau^4}$$

$$- \cos\left(\frac{\pi}{5} \right) \cdot \ln\left(\tau^4 - 2\tau^2 \cos\left(\frac{2\pi}{5} \right) + 1 \right)$$

$$- \cos\left(\frac{3\pi}{5} \right) \cdot \ln\left(\tau^4 + 2\tau^2 \cos\left(\frac{\pi}{5} \right) + 1 \right)$$

$$+ 2\sin\left(\frac{\pi}{5} \right) \left[\arctan\left(\frac{\tau - \cos\left(\frac{\pi}{5} \right)}{\sin\left(\frac{\pi}{5} \right)} \right) - \arctan\left(\frac{\tau + \cos\left(\frac{\pi}{5} \right)}{\sin\left(\frac{\pi}{5} \right)} \right) \right]$$

$$+ 2\sin\left(\frac{3\pi}{5} \right) \left[\arctan\left(\frac{\tau - \cos\left(\frac{3\pi}{5} \right)}{\sin\left(\frac{3\pi}{5} \right)} \right) - \arctan\left(\frac{\tau + \cos\left(\frac{3\pi}{5} \right)}{\sin\left(\frac{3\pi}{5} \right)} \right) \right]$$

$$+ \ln\left(\frac{5}{4} \right) + \cos\left(\frac{\pi}{5} \right) \cdot \ln\left(4 . \sin^2\left(\frac{\pi}{5} \right) \right) + \frac{3\pi}{5} \cdot \sin\left(\frac{\pi}{5} \right)$$

$$\left. + \cos\left(\frac{3\pi}{5} \right) \cdot \ln\left(4 . \sin^2\left(\frac{3\pi}{5} \right) \right) - \frac{\pi}{5} \cdot \sin\left(\frac{3\pi}{5} \right) \right\} \qquad (35)$$

The pure long range and short range order contributions are of particular interest. One gets from eqns (34) and (35)

$$\Delta H^{\mathrm{lro}}(T_\mathrm{c}) = 0 \cdot 645 . K^{\mathrm{lro}} . RT_\mathrm{c}$$

$$\Delta H^{\mathrm{sro}}(\infty) = 0 \cdot 598 . K^{\mathrm{sro}} . RT_\mathrm{c}$$

5.2 Determination of the Entropy of Ordering

In a similar way as for the enthalpy, the entropy can be derived from expressions (32) and (33) by integration. Here again the entropies of lro and sro will be evaluated separately:

$$\Delta S^{\text{lro}}(T) = \int_0^T \frac{c_p^{\text{lro}}}{T} \, dT \qquad \text{for } T < T_c$$

$$\Delta S^{\text{sro}}(T) = \int_{T_c}^T \frac{c_p^{\text{sro}}}{T} \, dT \qquad \text{for } T > T_c$$

These integrals can immediately be transformed to a series:[25]

$$\Delta S_{(T)}^{\text{lro}} = \frac{K^{\text{lro}}R}{3} \cdot F(\tau^3)$$

$$\Delta S_{(T)}^{\text{sro}} = \frac{K^{\text{sro}}R}{5} \cdot (F(1) - F(\tau^{-5}))$$

where

$$F(y) = \int_0^y x^{-1} \ln \frac{1+x}{1-x} \, dx = 2\left(y + \frac{y^3}{3^2} + \frac{y^5}{5^2} + \cdots \right) \qquad (36)$$

For $0 \le y \le \sqrt{2} - 1$ this series yields an accuracy better than 3×10^{-8} if the series is cut off at the sixth term. In order to get the same accuracy with six terms in the range $\sqrt{2} - 1 \le y \le 1$ one should make use of the following relation which $F(y)$ fulfills:

$$F(y) = \frac{\pi^2}{4} - \ln\left(\frac{1-y}{1+y}\right) \cdot \ln(y) - F\left(\frac{1-y}{1+y}\right)$$

In order to evaluate the total entropies of lro and sro one can use series (36) for $y = 1$:

$$F(1) = \frac{\pi^2}{4}$$

Thus one gets

$$\Delta S^{\text{lro}}(T_c) = S(T_c) - S(0K) = \frac{K^{\text{lro}}\pi^2 R}{12}$$

$$\Delta S^{\text{sro}}(\infty) = S(\infty) - S(T_c) = \frac{K^{\text{sro}}\pi^2 R}{20}$$

5.3 Determination of T_c and $\Delta S^{\text{ord} \to \text{dis}}$

With the formulae in Sections 5.1 and 5.2 the Gibbs energy of ordering depends on the two parameters T_c and $\Delta S^{\mathrm{ord}\to\mathrm{dis}}$. The critical temperature of lro has to be determined by a separate treatment based on statistical model calculations, e.g. references 16, 17 and 26. The entropy $\Delta S^{\mathrm{ord}\to\mathrm{dis}}$ is readily obtained in the case of atomic lro by the relation

$$\Delta S^{\mathrm{ord}\to\mathrm{dis}} = -R\left[\sum_I x_I \ln x_I - \sum_I \frac{1}{v}\sum_{L=1}^{v} p_I^L \ln p_I^L\right]$$

where p_I^L is the fraction of atoms I occupying a sublattice position L in the fully ordered state which is defined by the type of lro. Generally for bcc solutions four, and for fcc solutions eight, sublattices need to be considered,[26] ($v = 4$ or $v = 8$ respectively).

In the case of magnetic ordering one gets for the entropy difference between the paramagnetic (pm) and ferromagnetic state (fm) approximately[16,25]

$$\Delta S^{\mathrm{fm}\to\mathrm{pm}} = R\sum_I x_I \ln(\mu_I + 1)$$

where μ_I is the atomic moment of component I in Bohr magnetons.

APPENDIX I

Selected thermodynamic data for the phases of the Fe–V system (G, $^E H$ in J/mol; $^E S$ in J/K mol):

Liquid Phase

$$^E G^L(2193\,\mathrm{K}) = x_V x_{\mathrm{Fe}}[-\cdot180\,000E+05 - \cdot154\,800E+05(x_V - x_{\mathrm{Fe}})$$
$$-\cdot784\,500E+04(x_V - x_{\mathrm{Fe}})^2 + \cdot631\,800E+04(x_V - x_{\mathrm{Fe}})^3$$
$$+\cdot672\,000E+04(x_V - x_{\mathrm{Fe}})^4]$$

With $^E H^L = {}^E H^{\mathrm{bcc}}$ one obtains at $T = 2193\,\mathrm{K}$:

$$^E H^L = x_V x_{\mathrm{Fe}}[-\cdot228\,530E+05 - \cdot671\,532E+04(x_V - x_{\mathrm{Fe}})$$
$$-\cdot122\,131E+04(x_V - x_{\mathrm{Fe}})^2]$$

$$^E S^L = x_V x_{\mathrm{Fe}}[-\cdot244\,555E+01 + \cdot399\,703E+01(x_V - x_{\mathrm{Fe}})$$
$$+\cdot302\,038E+01(x_V - x_{\mathrm{Fe}})^2$$
$$-\cdot288\,091E+01(x_V - x_{\mathrm{Fe}})^3 - \cdot306\,407E+01(x_V - x_{\mathrm{Fe}})^4]$$

Bcc Phase

$$^EH^{bcc} = x_V x_{Fe}[-\cdot228\,530E+05 - \cdot671\,532E+04(x_V - x_{Fe})$$
$$- \cdot122\,131E+04(x_V - x_{Fe})^2]$$

$$^ES^{bcc} = x_V x_{Fe}[-\cdot658\,018E+01 + \cdot390\,813E+01(x_V - x_{Fe})$$
$$+ \cdot380\,540E+01(x_V - x_{Fe})^2$$
$$- \cdot217\,985E+01(x_V - x_{Fe})^3 - \cdot246\,840E+01(x_V - x_{Fe})^4]$$

Sigma Phase

$$\Delta H^{\alpha\rightarrow\sigma}_{FeV} = \cdot460\,328E+04 - \cdot323\,108E+05 \cdot x_V + \cdot336\,583E+05 \cdot x_V^2$$
$$\Delta S^{\alpha\rightarrow\sigma}_{FeV} = - \cdot222\,631E+00 - \cdot787\,789E+01 \cdot x_V + \cdot820\,614E+01 \cdot x_V^2$$

Combined with the data for the bcc phase and the lattice stabilities of the pure components according to eqn (1) these yield:

$$^EH^\sigma = x_V x_{Fe}[-\cdot565\,049E+05 - \cdot671\,532E+04(x_V - x_{Fe})$$
$$- \cdot122\,131E+04(x_V - x_{Fe})^2]$$

$$^ES^\sigma = x_V x_{Fe}[-\cdot147\,848E+02 + \cdot390\,813E+01(x_V - x_{Fe})$$
$$+ \cdot380\,540E+01(x_V - x_{Fe})^2$$
$$- \cdot217\,985+01(x_V - x_{Fe})^3 - \cdot246\,840E+01(x_V - x_{Fe})^4]$$

APPENDIX II

Lattice stability values used (J/mol):

$$V: \quad ^0G^L - {}^0G^{bcc} = \cdot182\,422E+05 - \cdot836\,800E+01 \cdot T$$
$$^0G^\sigma - {}^0G^{bcc} = \cdot594\,965E+04 - \cdot105\,604E+00 \cdot T$$
$$Fe: \quad ^0G^L - {}^0G^{bcc} = \cdot138\,072E+05 - \cdot763\,162E+01 \cdot T$$
$$^0G^\sigma - {}^0G^{bcc} = \cdot460\,328E+04 + \cdot222\,589E+00 \cdot T$$

REFERENCES

1. Kaufman, L. and Bernstein, H. (1970). *Computer Calculation of Phase Diagrams*, Academic Press, New York.
2. Hillert, M. (1978). In *Hardenability Concepts with Applications to Steel*, (Ed. D. V. Doane and J. S. Kirkaldy), The Metallurgical Society of AIME.
3. Ralston, A. and Wilf, H. S. (1967). *Mathematical Methods for Digital Computers*, Vol. 2, J. Wiley & Sons.

4. Fletcher, R. and Powell, M. F. D. (1963). *Computer Journal*, **6**(2), 163–8.
5. Fletcher, R. and Reeves, C. M. (1964). *Computer Journal*, **7**(2), 149–54.
6. Nelder, J. A. and Mead, R. (1965). *Computer Journal*, **7**, 308.
7. Counsell, J. F., Lees, E. B. and Spencer, P. J. (1971). *Met. Sci. J.*, **5**, 210.
8. Grimvall, G. and Ebbsjö, I. (1975). *Physica Scripta*, **12**, 168.
9. Darken, L. S. and Gurry, R. W. (1953). *Physical Chemistry of Metals*, McGraw-Hill, New York, p. 125.
10. Hansen, M. and Anderko, K. (1958). *Constitution of Binary Alloys*, McGraw-Hill, New York.
11. Reference 1, p. 74–8.
12. Reference 9, p. 275.
13. Reference 1, p. 79–91.
14. Boom, R., de Boer, F. R. and Miedema, A. R. (1976). *J. Less Common Metals*, **45**, 237; (1976). **46**, 271.
15. Miedema, A. R., Boom, R. and de Boer, F. R. (1975). *J. Less Common Metals*, **41**, 283; (1976). **46**, 67; (1975). 'Simple rules for alloying' in *Crystal Structure and Chemical Bonding in Inorganic Chemistry*, (Ed. C. J. M. Rooymans and A. Rabenau), North Holland Publ. Comp.
16. Inden, G. (1975). *Z. Metallkde.*, **66**, 577; (1975). **66**, 649; (1977). **68**, 529.
17. Inden, G. (1977). *J. de Physique*, **38**, Supp. C7-373.
18. Geiger, K. H., Probst, H. and Kubaschewski, O. (1977). *Z. Phys. Chemie*, **104**, 23.
19. Spencer, P. J. and Putland, F. H. (1973). *J. Iron and Steel Inst.*, **211**, 293.
20. Preston, R. S., Larn, D. J., Nevith, M. V., van Ostenburg, D. O. and Kimball, C. W. (1966). *Phys. Rev.*, **149**, 440.
21. Hack, K., Nüssler, D., Spencer, P. J. and Inden, G. (1979). Report of the Project Meeting CALPHAD VIII, Stockholm, p. 244.
22. Allibert, C., Bernard, C., Nüssler, D. and Spencer, P. J. (1979). Report of the Project Meeting CALPHAD VIII, Stockholm, p. 207.
23. Shao, J. and Machlin, E. S. (1978). *Calphad*, **2**, 279.
24. Inden, G. (1975). *Z. Metallkde.*, **66**, 577.
25. Inden, G. (1976). Report of the Project Meeting CALPHAD V, Düsseldorf, p. III.4-1.
26. Inden, G. (1974). *Acta Met.* **22**, 945; (1976). Report of the Project Meeting CALPHAD V, Düsseldorf, p. IV.1-1.

3

The Effect of Lattice Defects on Particle Generation, Distribution and Morphology

P. A. BEAVEN*

*Department of Metallurgy and Science of Materials,
University of Oxford, Oxford, UK*

1. INTRODUCTION

The decomposition of a supersaturated solid solution into one or more phases may generally be regarded as taking place in three stages:

(i)   the formation of nuclei of the new phase;

(ii)   growth of the nuclei to impingement or depletion of the matrix; and

(iii) coarsening, the process whereby the larger particles consume the small (Ostwald ripening) with the volume fraction of the particles remaining constant.

Consecutive or competing reactions may complicate this simple picture, and in reality this sequence of events should be regarded as a continuous process.

The most notable feature of solid–solid transformations of this type is the importance of kinetic factors. Diffusion rates in solids are generally orders-of-magnitude slower than diffusion rates in liquids, and this sluggishness may cause marked departures from predictions made from a strict thermodynamic analysis of phase diagrams. Equilibrium is eventually achieved via the path which maximises the *rate* of decrease of free energy of the system, rather than waiting for the necessary composition and crystallographic fluctuations which would yield the equilibrium phases directly, with a maximum decrease in the magnitude of the free energy of the system. The decomposition process may thus involve:

(i)   the formation of a series of metastable phases which progressively give rise to the formation of the stable products; and

(ii)   making use of the catalytic effect of lattice defects.

* Present address: Institut für Physik, GKSS Forschungszentrum, Geesthacht, FRG.

55

The interplay between thermodynamic and kinetic factors is a feature of all stages of the decomposition process. Although the major part of this lecture will be concerned with nucleation phenomena, the intrusion of growth and coarsening is, to some extent, unavoidable. This stems from the continuous nature of the process itself, and from the fact that the influence of lattice defects is both thermodynamic and kinetic in origin. References to growth and coarsening are thus only made where they are essential to the role of lattice defects in determining the distribution and morphology of the phases formed.

We begin by considering the principles of classical homogeneous nucleation theory and the modifications necessary for its application to solid state transformations. This will be followed by sections devoted to the role of lattice defects, classified as (i) point defects, (ii) line defects, and (iii) planar defects. In each case we seek the physical basis for the catalytic effect of the defect, as well as the influence on the kinetics of nucleation. The overall distribution of particles is then described in terms of the distribution at a particular defect and in terms of the distribution of defects. Finally we assess the influence of each class of defect on the nature and morphology of the phases formed.

It will become apparent that heterogeneous nucleation theory is considerably less quantitative than homogeneous nucleation theory, largely because of the difficulty in calculating the relevant nucleation parameters, together with the variety of nucleation mechanisms which may occur. The development of the capability to predict, even qualitatively, the microstructure formed during the course of a precipitation reaction, has relied on experimental observations to a great extent, and specific examples will be presented here to illustrate this approach.

2.   HOMOGENEOUS NUCLEATION THEORY

Nucleation theory was founded by Gibbs and has been developed extensively by chemists, physicists and metallurgists, for applications to vapour $\rightarrow$ liquid, vapour $\rightarrow$ solid, liquid $\rightarrow$ solid, and solid $\rightarrow$ solid transformations.[1]

Homogeneous nucleation may be defined as the process through which a fluctuation produces the smallest, stable aggregate of a new phase, without the assistance of a pre-existing heterogeneity.

The *driving force* for the nucleation is the volume free energy change, ΔG_v, associated with the transformation from α to β. The *barrier* to

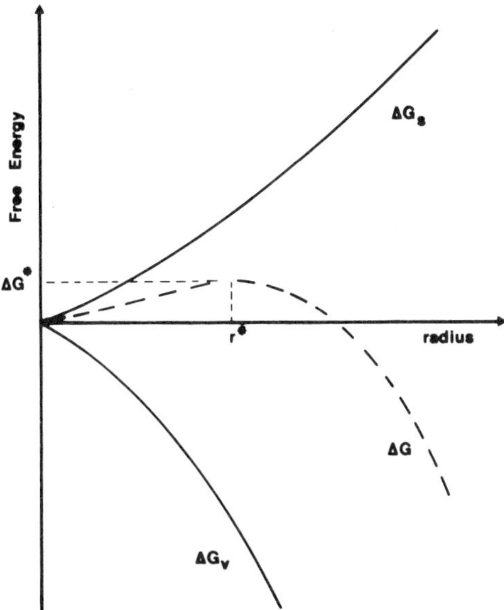

FIG. 1.   Free energy versus radius of nucleus.

nucleation is the interfacial free energy of the nucleus–matrix interface, ΔG_s. As shown in Fig. 1, the free energy of formation of a spherical nucleus, ΔG has two components, ΔG_v which is negative and proportional to nucleus volume (r^3), and ΔG_s, which is positive and proportional to the surface area of the nucleus (r^2). When r is very small, ΔG_s increases more rapidly than ΔG_v; the reverse situation occurs when r is relatively large.

When $\partial(\Delta G)/\partial r = 0$, the radius of the nucleus is referred to as r^*, the *critical radius*. Nuclei which attain this critical size are then able to grow further with a reduction in free energy. The value of ΔG corresponding to r^*, is referred to as ΔG^*, *the free energy of formation of the critical nucleus*. For spherical nuclei, with interfacial energy σ,

$$\Delta G = -\Delta G_v + \Delta G_s \tag{1}$$

$$\Delta G = -4/3 \, . \, \pi r^3 \, . \, \Delta G_v + 4\pi r^2 \sigma$$

$$\frac{\partial(\Delta G)}{\partial r} = -4\pi r^2 \, . \, \Delta G_v + 8\pi r \sigma = 0$$

Thus

$$r^* = \frac{2\sigma}{\Delta G_v} \tag{2}$$

and substitution gives

$$\Delta G^* = \frac{16\pi\sigma^3}{3\Delta G_v^2} \tag{3}$$

(Note the importance of the interfacial energy term.)

The application of this analysis to the formation of nuclei in the solid state requires the incorporation of an additional energy term which results from the difference in specific volume between nucleus and matrix. Since the volume change attending the transformation cannot easily be accommodated, this serves to increase the nucleation barrier. Thus

$$\Delta G = (-\Delta G_v + \Delta G_{\text{Strain}}) + \Delta G_s \tag{4}$$

since ΔG_{Strain} is proportional to nucleus volume. As a result, eqns (2) and (3) now become

$$r^* = \frac{2\sigma}{\Delta G_v + \Delta G_{\text{Strain}}} \tag{5}$$

and

$$\Delta G^* = \frac{16\pi\sigma^3}{3(\Delta G_v + \Delta G_{\text{Strain}})^2} \tag{6}$$

The necessary condition for a successful nucleation event is that r^* be achieved, and this is affected by the balance between the various energy terms in eqns (5) and (6). The nucleation process is therefore strongly influenced by the particular compositions, crystal structures and elastic constants of the matrix and precipitate phases, and we now consider these parameters:

ΔG_s

The interfacial energy, σ, between matrix and precipitate is a strong function of the nature of the interphase interface. It is composed of a chemical energy term which reflects the composition change across the interface, and a structural term which is a measure of the crystallographic matching across the interface. If both phases are of identical or similar crystal structures, there is the possibility that the precipitate will nucleate preferentially on certain crystallographic planes and along specific directions, producing interfaces across which the lattice matching is perfect. The interface is characterised by low energy and is termed *coherent*.

If the crystal structures of precipitate and matrix are quite different, good lattice matching may be impossible; the interface is thus disordered at all

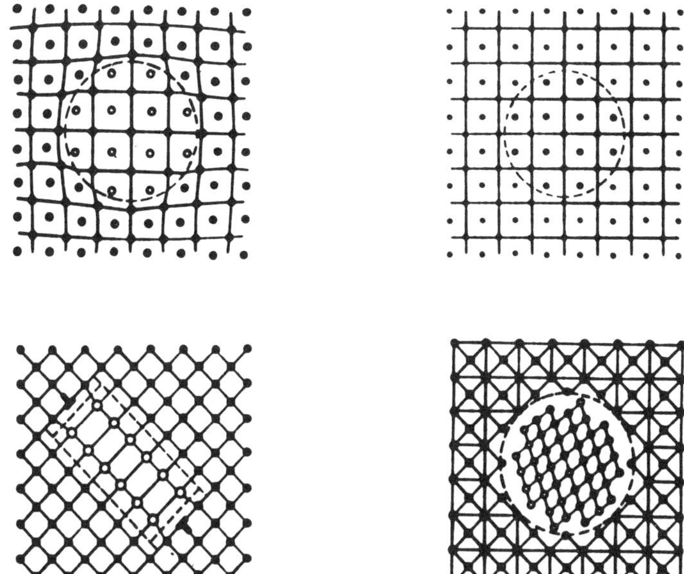

Fig. 2. Schematic diagrams of various interphase interfaces. Top left: coherent with misfit; top right: coherent with zero misfit; bottom left: semicoherent with interfacial dislocation at edge, coherency at broad face; bottom right: incoherent.

orientations and σ is isotropic and of a high value. Such an interface is termed *incoherent*.

When only a limited amount of lattice matching is possible and the precipitate–matrix interface is made up of regions of good fit interspersed with regions of bad fit, the interface is described as *semicoherent*, and is characterised by an intermediate value of interfacial energy. In this case, the structural component of the surface energy is readily seen to result from the network of dislocations which lie in the interface. Schematic diagrams of these types of interface are shown in Fig. 2.

Since ΔG^* is a function of σ^3, the barrier to homogeneous, incoherent, nucleation is extremely high. Reductions in ΔG^*, and r^* may thus be achieved for semicoherent, and coherent nucleation respectively. In the general case, the crystallographic matching between phases will depend on their relative orientations, and particles will be bounded by some interfaces of low energy, and some of high energy. Solid–solid surface energies are highly anisotropic, and the preferred nucleus shape will tend to be that which minimises the surface energy and not simply the surface area.

ΔG_{Strain}

The strain energy term arises from differences in specific volume between precipitate and matrix. For the case where precipitate and matrix have identical crystal structures, but different lattice parameters, the mismatch, δ, may be written as

$$\delta = \frac{a_\alpha - a_\beta}{a_\beta} \tag{7}$$

where a_α is the lattice parameter of the matrix, and a_β is the lattice parameter of the precipitate. We now consider the strain energy term as a function of δ.

(1) *For $\delta = 0$,* i.e. $a_\alpha = a_\beta$; no strain energy would be associated with the nucleation event and the nucleus would be perfectly coherent.

(2) *For $\delta < 0{\cdot}05$;* the mismatch at the interface may be taken up by elastic strains in the lattices and the interface remains coherent. The strain energy per unit volume of nucleus is proportional to δ^2 and may be calculated using equations derived by Eshelby[2] assuming isotropic elasticity theory. The strain energy per unit volume is found to be independent of particle shape, if matrix and nucleus have the same elastic constants ($\mu_\alpha = \mu_\beta$). For the case of an incompressible matrix, the elastic strain is taken up by the nucleus and is approximately three times greater. For the general case where $\mu_\beta > \mu_\alpha$, most of the strain energy is taken up by the matrix, and due to the anisotropy of the elastic constants, the strain energy associated with a coherent nucleus, may be reduced if nucleation takes place on crystallographic planes normal to the elastically 'soft' directions. In cubic lattices, coherent precipitates thus tend to form as plate-shaped particles on $\{100\}$ planes, because the modulus of elasticity is a minimum along $\langle 100 \rangle$ directions.

(3) *For $0{\cdot}05 < \delta < 0{\cdot}25$;* the elastic strain energy may be relieved by the formation of a network of dislocations in the interface, yielding a semicoherent boundary. Thus ΔG_{Strain} may be reduced, but only at the expense of raising the structural component of the interfacial energy ΔG_s. Brown *et al.*[3] have attempted to calculate the strain energy change which occurs when a dislocation loop is formed at a precipitate–matrix interface, but application of this to small nuclei would serve only as a rough approximation. It is possible that a greater reduction in energy could be obtained by a change in nucleus shape.[4]

(4) *For $\delta > 0{\cdot}25$;* the interface between the phases may be regarded as

incoherent, and the only elastic strains present are those arising from the different specific volumes of the phases. The strain energy is shape dependent,[5] being a maximum for spherical particles, although for such a shape (minimum surface area) the interfacial energy term is minimised. The high surface energy term is likely to control the nucleus shape in this case, and thus the homogeneous nucleation of incoherent particles is unlikely to occur.

For the more general case, where matrix and nucleus do not possess a parallel orientation relationship, δ will vary with orientation, and the nucleus will tend to adopt a shape which minimises the total strain and surface energies.

ΔG_v

In a multicomponent system, ΔG_v is composition and temperature dependent with respect to the overall concentration of solute in the matrix and the composition of the precipitate particle. We will be concerned with portions of a typical phase diagram as shown in Fig. 3. For the decomposition of a thermodynamically ideal alloy of initial composition C, and of equilibrium solute concentration, C_0, at a temperature T,

$$\Delta G_v = - RT \ln \frac{C}{C_0} \tag{8}$$

where R is the gas constant.

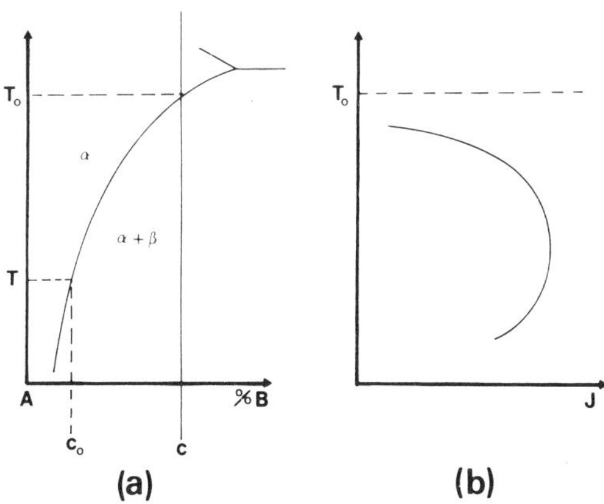

FIG. 3.   (a) Binary phase diagram showing alloy of composition C, considered in text. (b) Variation of nucleation rate J, with temperature.

62   P. A. BEAVEN

Since C_0 is exponentially related to the temperature via the heat of solution Q,

$$\frac{C}{C_0} = \exp\left(\frac{-Q\,\Delta T}{RT^2}\right) \qquad (9)$$

where ΔT is the supercooling below the solvus temperature.

For $\Delta T \ll T$,

$$\Delta G_v = \frac{Q \cdot \Delta T}{T} \qquad (10)$$

The magnitude of ΔG_v may therefore be described, either by the undercooling ΔT, or in terms of the supersaturation ratio C/C_0.

2.1 Kinetics of Nucleus Formation

The steady state nucleation rate J^* (i.e. the number of nuclei forming per unit volume per unit time) may be written as[4]

$$J^* = Z \cdot \beta^* \cdot C^* \qquad (11)$$

where Z is the Zeldovich factor, β^* is the rate at which atoms join the critical nucleus, and C^* is the concentration of critical nuclei.

The concentration of critical nuclei is given by[4]

$$C^* = \frac{N_s}{\bar{C}} \exp - (\Delta G^*/kT) \qquad (12)$$

where N_s is the density of potential nucleation sites, $\bar{C}$ is the composition of the nucleus, and ΔG^* is defined by eqn (6). The rate at which individual nuclei grow will also be a function of the diffusion rate which is known to vary exponentially with temperature. Thus the rate of precipitate nucleation may be written as

$$J = K \exp - \left(\frac{\Delta G^* + \Delta G_D}{kT}\right) \qquad (13)$$

where ΔG_D is the activation energy for diffusion. The precipitate nucleation rate varies with temperature as shown in Fig. 3(b), where at small undercoolings the rate is thermodynamically controlled, and at large undercoolings, the rate is kinetically controlled. (Note also that a critical undercooling must be exceeded in order to achieve a sensible rate of nucleation.)

2.2 Conclusions

The magnitude of ΔG^* is controlled by the balance of the energy terms $\Delta G_v + \Delta G_s$, of which ΔG_s plays the most significant role in the earliest stages. The steady-state nucleation rate is controlled by the value of ΔG^*. The theory thus predicts that homogeneous nucleation is only likely for coherent (minimum σ) nuclei, with small values of δ (low ΔG_{Strain}). For such special cases the various parameters may be calculated with a reasonable degree of precision, allowing comparisons to be made between predicted and observed nucleation rates. This has been achieved for Cu–Co alloys[6] and for Ni–Al alloys,[7] and satisfactory *quantitative* agreement has been found.

For the more general case of a typical equilibrium precipitate, the supercooling necessary for a sensible rate of nucleation has been estimated by Nicholson[8] to be of the order of 300–400 K for a solvus temperature of 1000 K. The diffusivity will thus be too small to allow nucleation to occur within a reasonable time. With few exceptions, nucleation in the solid state is either heterogeneous or is accomplished via a homogeneously nucleated transition precipitate.

We now briefly consider the latter possibility before proceeding to heterogeneous nucleation. Figure 4 shows a schematic phase diagram of a system in which the formation of metastable, coherent or semicoherent, intermediate phases may be expected during ageing. With increasing

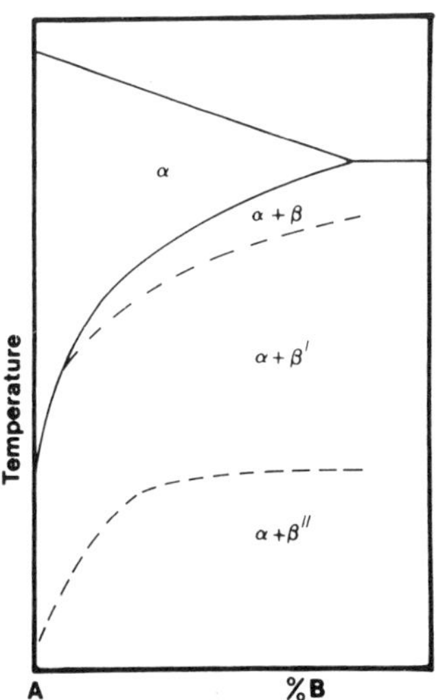

FIG. 4.   Schematic phase diagram showing solvi for metastable β', β'' phases.

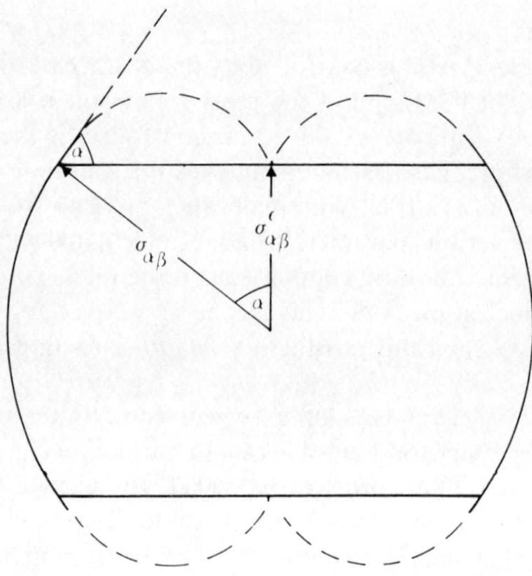

FIG. 5.   Facetted nucleus shape utilised by Aaronson and Lee.[9]

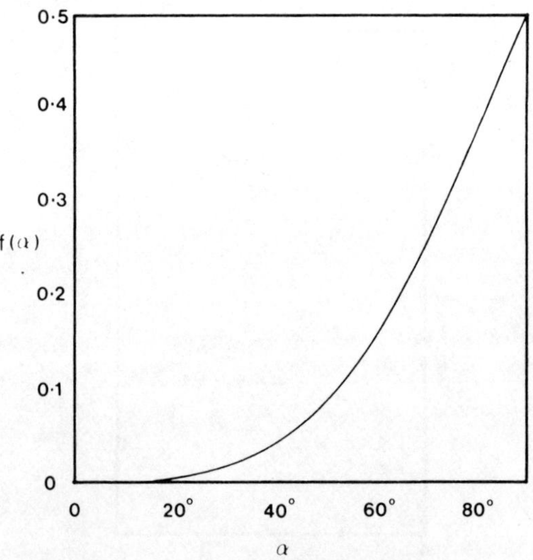

FIG. 6. $f(\alpha)$ as a function of α.[9]

undercooling, the number of transition precipitates involved in the decomposition process increases, i.e. the full sequence would involve

$$\alpha \to \alpha + \beta'' \to \alpha + \beta' \to \alpha + \beta$$

We note that ΔG_v for the metastable phases is always less than ΔG_v for the equilibrium β phase. For ΔG^* to be minimised, the most important factor must be the potential reduction in σ which may be achieved by the structural similarities between transition phase and matrix (even at the expense of raising ΔG_{Strain}).

The formal calculation of ΔG^* in the general case requires the simultaneous minimisation of ΔG_v, ΔG_s and ΔG_{Strain}, for the assumed critical nucleus shape. Critical nuclei are so small that details of their shape and composition are difficult to obtain. It is thus necessary to assume that both shape and composition are characteristic of equilibrium. (The composition dependence of these energy terms is not yet understood quantitatively.) Lee and Aaronson[9] have attempted to minimise ΔG^* for a facetted spherical nucleus, by placing primary emphasis upon minimisation of the interfacial energy. The errors introduced by ignoring the effects on ΔG_{Strain} may not be appreciable since the interfacial energy is raised to the third power in eqn (6), and are insignificant, if the volume change attending nucleation is small. The nucleus shape is characterised by a low energy interface at one orientation of the α/β boundary and is of the form shown in Fig. 5. Assuming $\Delta G_{\text{Strain}} = 0$,

$$\Delta G = -\Delta G_v . 2\pi . r^3 \left(\cos\alpha - \frac{\cos^3\alpha}{3} \right) + 2\sigma_{\alpha\beta}^c . \pi r^2 . \sin^2\alpha + 4\sigma_{\alpha\beta} . \pi r^2 \cos\alpha$$

$$(14)$$

Where the first term represents the product of ΔG_v and nucleus volume, the second term gives the total interfacial energy of the two parallel facets (where $\sigma_{\alpha\beta}^c$ is the coherent interfacial energy), and the third term accounts for the interfacial energy of the spherically curved portion of the interface.

It is found that

$$\Delta G_{\text{hf}}^* = \Delta G_{\text{hu}}^*(1 - 2f . (\alpha)) \tag{15}$$

where subscript hf represents homogeneous facetted, subscript hu refers to homogeneous unfacetted, and $f(\alpha) = \frac{1}{4}(2 - 3\cos\alpha + \cos^3\alpha)$. $f(\alpha)$ as a function of α is shown in Fig. 6. The potential reduction in ΔG^* is thus governed by the value of α, which is itself a function of the ratio $\sigma_{\alpha\beta}^c/\sigma_{\alpha\beta}$. Homogeneous nucleation therefore becomes a much more likely possibility if $\sigma_{\alpha\beta}^c/\sigma_{\alpha\beta}$ is sufficiently small.

In the case of aluminium alloys, that such nucleation is truly homogeneous is perhaps questionable due to the role of quenched-in vacancies (to be discussed later). In terms of the distribution of precipitates formed, there is abundant evidence to show that the nucleation of successive transition precipitates in the sequence is influenced by the distribution of their predecessors, although the mechanisms operating are not well understood.[10] However, low temperature ageing of such alloys to produce fine dispersions ($\sim 10^{18}$ cm^{-3}) of transition precipitates may result in the production of finer dispersions of successive precipitates and of the equilibrium phase.

3. HETEROGENEOUS NUCLEATION AT LATTICE DEFECTS

The influence of lattice defects on nucleation in the solid state derives from their ability to alter the balance of the *thermodynamic* terms in eqn (4), and their influence on the *kinetics* of nucleation by providing paths of enhanced diffusivity. Lattice defects which are not in equilibrium correspond to high energy regions, and when the defects are destroyed or partially destroyed during the nucleation event, the energy released is able to contribute to reducing the energy barrier ΔG^*.

The steady-state *heterogeneous* nucleation rate may be written as

$$J^*_{het} = Z \cdot \beta^*_{het} \frac{N_s}{\bar{C}} \exp\left(\frac{-\Delta G^*_{het}}{kT}\right) \tag{16}$$

This may be compared with the similar equation (eqn (6)) for homogeneous nucleation.

The ratio of nucleation rates may be written as

$$\frac{J^*_{het}}{J^*_{hom}} \simeq \frac{N^{het}_s}{N^{hom}_s} \cdot \exp - \left(\frac{\Delta G^*_{het} - \Delta G^*_{hom}}{KT}\right) \tag{17}$$

since Z, β^*, $\bar{C}$ are not changed markedly in the two cases. Although $\Delta G^*_{het} < \Delta G^*_{hom}$, it is important to realise that $N^{het}_s \ll N^{hom}_s$, and it is this factor which prevents the complete dominance of heterogeneous nucleation.

We now consider the role of specific defects and classify these as follows:

(1)   point defects,
(2)   line defects—dislocations,
(3)   planar defects—grain boundaries.

The following points are considered for the various classes of defect:

(i)   the physical models for the interaction between defect and nucleus;
(ii) the influence on precipitate distribution;
(iii) the influence on precipitate morphology;
(iv) the influence on the nature of the precipitate.

(The effect of defects is often great enough to allow direct formation of the equilibrium phase, while precipitation within the matrix may proceed by the homogeneous nucleation of a metastable transition phase.)

3.1. Point Defects

Many precipitation hardening alloys are quenched from a high temperature and aged at lower temperatures. Since the equilibrium concentration of vacancies rises exponentially with temperature, an excess vacancy concentration at the ageing temperature may be 'frozen in' during the quench. During ageing these excess vacancies may assist the nucleation and growth of precipitates. Because single vacancies and clusters of vacancies are unobservable in the transmission electron microscope, much of the evidence obtained is of an indirect nature.

The presence of excess vacancy concentrations may have the following effects:

(i)   For precipitates with larger specific volumes than the matrix, vacancies may enter the precipitate thus reducing the value of ΔG_{Strain} with a corresponding reduction in ΔG^*.
(ii) Vacancies may aggregate together to form a variety of lattice defects, e.g. dislocation loops.
(iii) The presence of vacancies may enhance the diffusion rate of solute atoms.
(iv) The migration of vacancies to lattice defects such as dislocations and grain boundaries may induce non-equilibrium segregation of solute atoms to these sites.

In this section we consider points (i) and (iii) and defer treatment of (ii) and (iv) until the sections on dislocations and grain boundaries.

3.1.1 Effect of Vacancies on Precipitate Distribution

Consider a precipitate nucleus with a value of δ, such that $\Delta G_{\text{Strain}} > -\Delta G_v$. Thus $\Delta G > 0$, and nucleation cannot, in principle, take place. If excess vacancies form an integral part of the precipitate nucleus, (1) the vacancies modify δ, and (2) there is a decrease in free energy, (Δg_v), due to the matrix vacancy concentration approaching its equilibrium value.

The number of vacancies included in the nucleus is limited and may be expressed as

$$\sum_r N_r \cdot \rho \cdot \frac{C_n r^3}{a_3} = n - n_e$$

where N_r is the number of nuclei of radius r, ρ is the number of vacancies per atom, C_n is the number of atoms per unit cell, a is the lattice parameter, and n and n_e are the number of vacancies after quenching and at equilibrium respectively.

For strain-free nucleation ($\delta = 0$) the nucleus requires ρ_0 vacancies per atom where

$$\rho_0 = \frac{V^\beta - V^\alpha}{V^\alpha}$$

where V is the volume per atom. Thus, in the presence of vacancies, the strain energy function is modified by a factor

$$\left(1 - \frac{\rho}{\rho_0}\right)$$

The additional driving force, Δg_v, for the 'precipitation' of vacancies is

$$\Delta g_v = -kT \cdot \ln \cdot \frac{n}{n_e}$$

Thus in eqn (6), the term $(-\Delta G_v + \Delta G_{Strain})$ may be replaced by $-\Delta G_v + \Delta G_v + \Delta G_{Strain}(1 - \rho/\rho_0)$. The parameter which thus determines whether nucleation will occur is the value of ρ/ρ_0 which relates to the excess vacancy concentration.

This model has been used successfully to describe the influence of excess vacancy concentrations on the precipitation of niobium carbide in a 20%Cr–25%Ni austenitic stainless steel.[11] The niobium carbide has the same crystal structure as the matrix (fcc) but a lattice parameter which is greater by $\sim 24\%$. In such a case the barrier to coherent nucleation is dominated by the strain energy term.

MC carbide may form coherently with the matrix in other austenitic steels and cobalt-based alloys; indirect experiments have indicated that heat treatments which increase the excess vacancy concentration, increase the observed precipitate nucleation rate, i.e. the number density of particles is increased. Control of the distribution may be effected by; (1) the choice of solution treatment temperatures, (2) the control of cooling rate, and (3) irradiation treatments which produce high vacancy concentrations.

The distribution of particles is also influenced by the presence of other defects such as dislocations and grain boundaries, which act as sinks for vacancies. Thus particle-free regions are commonly observed adjacent to these defects.

There remains the possibility that particle nucleation actually occurs on small dislocation loops formed by the aggregation of vacancies. However in systems where this effect has been observed it is usually possible to select heat treatments in which these features are observable.

3.1.2 *Effect of Vacancies on Precipitation Kinetics*

The role of vacancies in accelerating solid state diffusion is best illustrated by studies of the kinetics of formation of precipitates or G–P zones in quenched aluminium alloys. The work in this field is extensive and has been reviewed by Herman.[12] Diffusivities of 10^{13} greater than the equilibrium value have been observed, and the interpretation of this is based on the assumption that solute atom–vacancy binding occurs. This assumption thus allows for increased rates of transport as well as the necessary prolongation of the vacancy life time. It is also a necessary consequence of this model that the vacancy does not enter the nucleus and becomes available to transport further solute atoms for growth to continue.

3.2 Line Defects

The simplest model for the nucleation of particles on dislocation lines is due to Cahn.[13] The formation of a precipitate nucleus is assumed to destroy the portion of dislocation line on which it is situated, thus releasing the strain energy of the dislocation to contribute to ΔG. Thus eqn (4) is modified to include an energy term relating to the dislocation strain energy.

$$\Delta G = (-\Delta G_v + \Delta G_{\text{Strain}}) + \Delta G_s - \Delta G_{\text{Dislocation}}$$

Cahn assumed a cylindrical nucleus with a spherical bulge. Furthermore he assumed an incoherent nucleus with zero misfit, i.e. $\Delta G_{\text{Strain}} = 0$. The reduction in nucleation barrier with respect to homogeneous nucleation for edge dislocations is given by

$$\frac{\Delta G^*_{\text{dislocation}}}{\Delta G^*_{\text{hom}}} \simeq 1 + \frac{\mu \mathbf{b}^2 \Delta G_v}{2\pi^2 . \sigma^2 (1-v)} \tag{18}$$

and for screw dislocations by

$$\frac{\Delta G^*_{\text{dislocation}}}{\Delta G^*_{\text{hom}}} \simeq 1 + \frac{\mu \mathbf{b}^2 . \Delta G_v}{2\pi^2 . \sigma^2} \tag{19}$$

where $\mathbf{b}$ is the Burgers vector, μ the shear modulus, and ΔG_v is of course negative. For

$$\frac{\mu \mathbf{b}^2 . \Delta G_v}{2\pi^2 . \sigma^2} > 1$$

the nucleation barrier disappears altogether and nucleation may occur spontaneously, i.e. as rapidly as diffusion will allow.

The theory predicts that:

(i)   nucleation is more favoured at edge dislocations than screw dislocations;

(ii)   nucleation is more favoured the larger the Burgers vector, and the larger the value of μ;

(iii)   nucleation will be most favoured when ΔG_v is more negative, i.e. at higher supersaturations.

Several other workers[14–17] have attempted more sophisticated calculations but the elasticity problem is complex and many assumptions are required, in particular about the state of coherency of the nucleus. A major limitation is the assumption of isotropic elasticity. The conclusions which result may be summarised as follows:

(i)   Coherent nuclei may form in regions close to a dislocation due to interactions between the long range strain fields of both nucleus and dislocation.[14]

(ii)   The reduction in nucleation barrier for coherent and incoherent nuclei is comparable.[18]

(iii)   For incoherent nuclei, reductions in supersaturation, or increased specific volume differences between nucleus and matrix, may lead to plate-like morphologies.[17]

(iv)   For incoherent nuclei, increases in supersaturation should lead to nucleation on both edge and screw dislocations, and also homogeneously in the matrix.[15]

This latter prediction is more appropriate to experimental observations than that of Cahn.[13] It is generally observed that precipitation on dislocations is dominant at low supersaturations, where the growth of dislocation-nucleated precipitates is sufficiently rapid to deplete the matrix of solute before homogeneous nucleation can occur. Evidence for the prediction concerning the magnitude of the Burgers vector of the dislocation has been reviewed elsewhere.[19] With regard to the state of coherency of dislocation-nucleated particles most observations indicate that coherent or semicoherent phases are often formed at dislocations, e.g. transition precipitates. This is not surprising, since the role of the dislocation involves reductions in ΔG_{Strain}, rather than ΔG_s. As a consequence of this, the influence of dislocations on nucleus morphology is usually minimal.

One feature of precipitation on dislocations, which is not accounted for by the theories, is that a given dislocation does not constitute a nucleation site of equal potency for all possible orientations of the precipitate. For plates of θ' forming on {100} planes in Al–Cu alloys, only two of the three possible {100} habits are observed at a given dislocation of Burgers vector $a/2 \langle 100 \rangle$.[10] This, and other observations, has led to the formulation of an empirical rule, which states that the nucleation potency is greatest for the precipitate variant, which has its maximum misfit vector very nearly parallel to the Burgers vector of the dislocation.[10] This appears to hold adequately for face-centred cubic alloys, but cannot be applied to body-centred cubic alloys, with plate-shaped particles on {100}. Observations of α'' precipitation in bcc Fe–N alloys,[20] show that a single α'' variant is formed at a given dislocation. In this case, the major misfit vectors (along $\langle 100 \rangle$ directions) are equally inclined to the Burgers vector $a/2 \langle 111 \rangle$. (Thus all three variants should be equally preferred.) Observations of α'' formation at dislocations of mixed, edge, and screw character, suggest that the dislocation line orientation with respect to the habit plane of the precipitate is an important factor (Fig. 7). Further work, both theoretical and experimental, is necessary to substantiate these effects, and the observation that dislocation movement may take place during nucleation (see later) may also be a significant factor, influencing the distribution and orientation of particles at individual dislocations.

3.2.1 Distribution of Dislocations and Particles

Useful densities of precipitates by nucleation on dislocations may be obtained by two general routes. The first, is to increase the density of dislocations by:

(i)   cold work prior to ageing;
(ii) martensitic transformation of the matrix;
(iii) the aggregation of point defects to produce dislocation loops.

The chief problem is to produce a *uniform* dislocation distribution. Dislocations produced by (i) are typically arranged in a cell structure with a cell diameter of $1 \mu m$, and the resulting distribution of precipitates is also non-uniform. The uniform distribution of carbide particles in certain

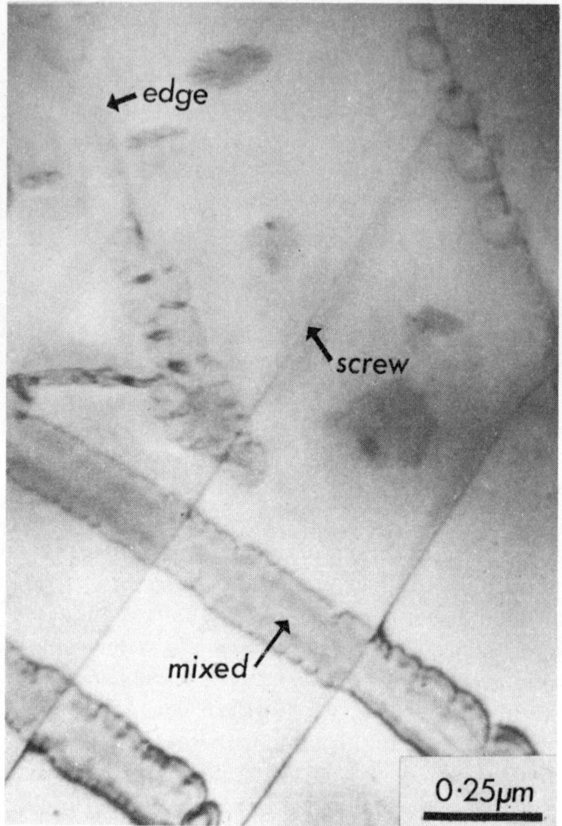

Fig. 7.   Precipitation of α'' on dislocations in Fe–N alloys showing preference for edge, and mixed dislocations.

martensitic steels is an example of (ii), and further refinement of the carbide dispersion may be achieved by warm working, prior to the martensitic transformation,[21] (the ausforming process). The large concentrations of vacancies required for (iii) may be introduced by quenching from high temperatures, or by irradiation. The conditions under which uniform distributions of vacancy loops are found are not well understood, although this does occur in some alloys. There is evidence that certain trace element additions may modify vacancy behaviour and thus exert an influence on the resulting precipitate dispersion, (e.g. phosphorus, boron in austenitic stainless steels).[22,23] See Fig. 8.

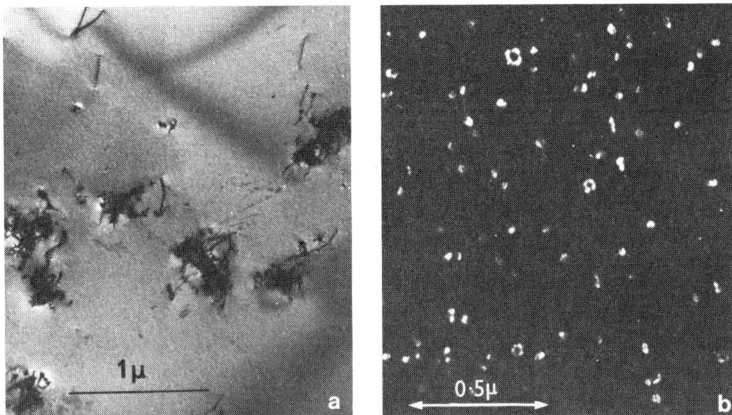

FIG. 8. Distribution of TiC particles in 18 Cr–12 Ni stainless steel arising from; (a) nucleation on dislocations (boron-free alloy), and (b) nucleation at vacancy loops (140 ppm boron added).

The second general way in which to utilise nucleation on dislocations, is to ensure that single dislocations are able to nucleate a series of precipitates over a period of time. This requires that the dislocation should move through the matrix into regions which are still supersaturated with respect to solute. This process, repeated nucleation on dislocations, occurs in a number of alloys. The basic mechanism requires that dislocations move by *climb* during the precipitation process; this may occur by vacancy absorption or by vacancy emission. The former has been observed during θ' precipitation in Al–Cu alloys, where the high vacancy concentrations obtained on quenching are *absorbed* at matrix dislocations or at vacancy loops, giving rise to colonies of θ' precipitates, Fig. 9.[24] (NB the θ' precipitate has a smaller specific volume than the matrix.) The other process occurs in a number of alloys where the specific volume of the precipitate is larger than the matrix. Although nucleation occurs at dislocations, continued particle growth may only occur with the aid of vacancies, and the dislocations thus climb in order to *create* vacancies which are then absorbed by the growing particles. Again, colonies of particles are formed by repeated nucleation at matrix dislocations, (for example, carbides in austenitic steels,[25] cobalt-based alloys,[26] and Cu–Ag alloys[27]). The most striking example, however, is the so-called 'stacking fault precipitation' which also occurs in similar alloys. Here the normal $a/2 \langle 100 \rangle$ matrix dislocations dissociate in such a manner as to form a stacking fault bounded

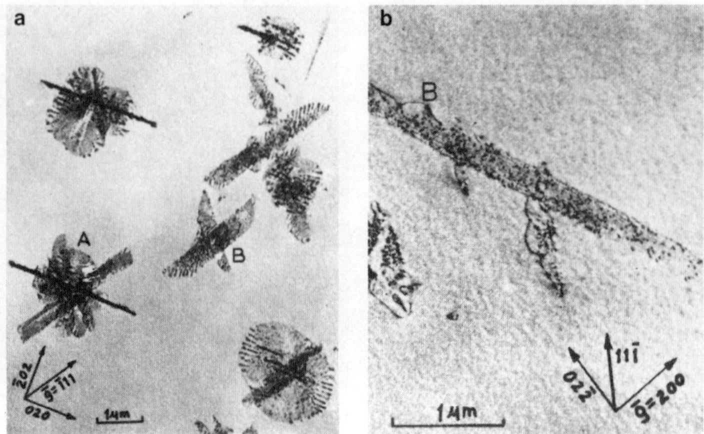

FIG. 9.   Repeated nucleation of θ' particles on; (a) vacancy dislocation loops, and (b) climbing lattice dislocations in an Al–Cu alloy.[24] Courtesy of *Phil. Mag.*

by a Frank partial dislocation of Burgers vector $a/3 \langle 111 \rangle$. It is this partial dislocation which climbs to produce vacancies, thus enlarging the stacking fault. During this climb process, repeated nucleation of precipitates occurs on the Frank partial dislocation, and the final microstructure consists of a series of disc-like stacking faults on {111} planes, several microns in diameter, each containing a fine distribution of particles (see Figs. 10 and 12). The most important requirement for this mechanism to operate appears to be a large precipitate–matrix mismatch, coupled with a similarity in crystal structure between precipitate and matrix. As a result, the number of alloys which show this phenomenon is rather limited.

These observations of repeated nucleation show clearly that dislocation motion has occurred during the precipitation process, and demonstrate the combined effects of vacancies and dislocations. For the nucleation and growth of a single set of precipitates at a dislocation, however, the usual experimental techniques of observation, would not enable us to establish whether dislocation movement accompanied the precipitation event. Recent experiments[28] using *in situ* transmission electron microscopy, have shown that during the nucleation and growth of β' precipitates in an Al–Mg–Si alloy, dislocation climb by absorption of excess vacancies, does in fact occur. The needle-shaped β' particles remain attached to the dislocation line, (Fig. 11) and the growth rate is higher in the direction of climb, than in the reverse direction, (Fig. 12). As far as nucleation is concerned, it appears that the dislocation climb places further restrictions

Fig. 10.   Precipitation of TiC particles by repeated nucleation on Frank partial dislocations in an 18 Cr–12 Ni austenitic stainless steel.

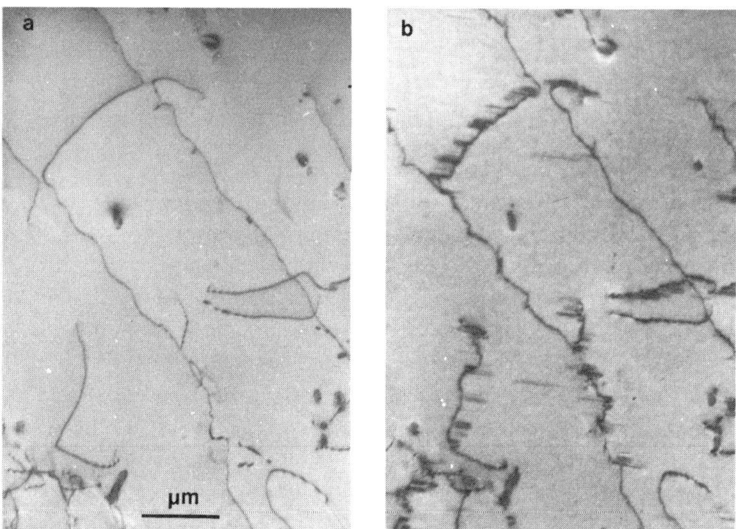

Fig. 11.   Nucleation and growth of β' needles at climbing dislocations in an Al–Mg–Si alloy; (a) before ageing, and (b) during *in situ* ageing in the high voltage electron microscope.

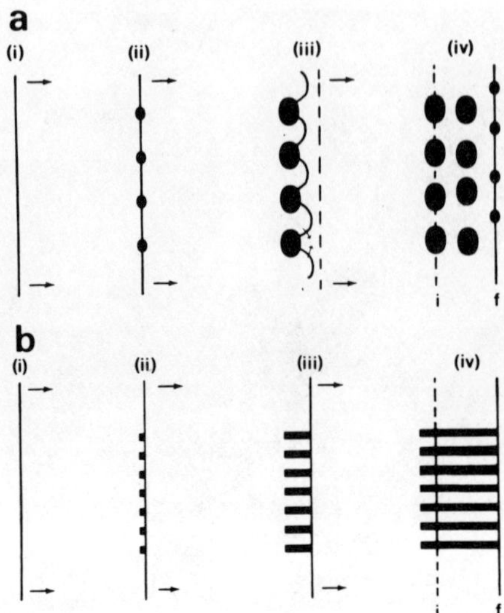

FIG. 12. Schematic diagram illustrating; (a) repeated nucleation at moving dislocations, and (b) particle growth in association with moving dislocations. i and f refer to initial and final positions of dislocations.

on the number of precipitate habits which may form. It is possible that this phenomenum occurs in many other alloys, and may be of relevance in high temperature creep.

3.2.2 Effect of Dislocations on Precipitation Kinetics

It is well known that 'pipe' diffusion along dislocations takes place with an activation energy which is approximately half that for volume diffusion. It has been argued[16] that this, together with an enhanced concentration of solute atoms arising from segregation, should influence the rate at which atoms are able to join the critical nucleus, and hence should increase J^*_{het}, particularly at low temperatures. (The segregation may be of the equilibrium, or non-equilibrium type as referred to in Section 3.1). For the growth of precipitates at dislocation lines the chief problem is to assess the relative contributions of volume diffusion to the particles, and pipe diffusion along the dislocations. Theoretical analysis has therefore been confined to the case of coarsening of particles on dislocations, with the assumption that diffusion along the 'pipe' is dominant. Here, a dependence of particle radius on $t^{1/5}$ is predicted,[29] and Fig. 13 shows an example where this has been confirmed experimentally.[30]

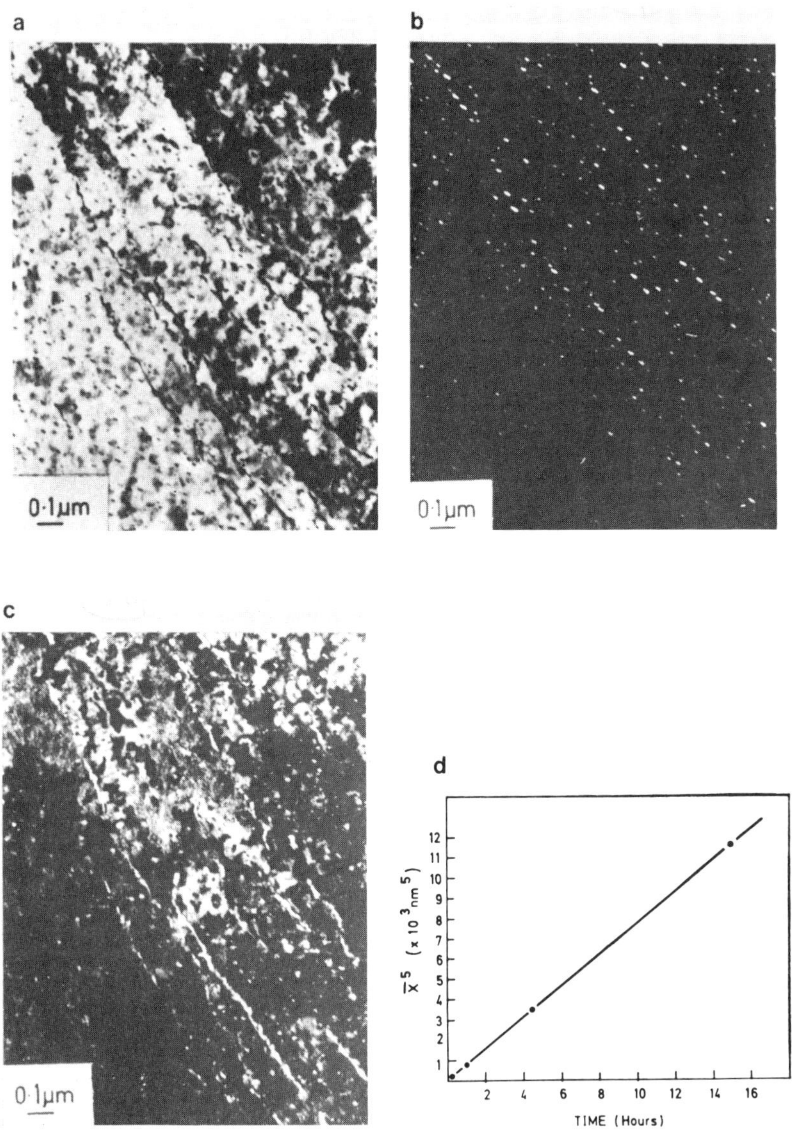

Fig. 13. Coarsening of TiC particles on dislocations in an Fe–Ti–C alloy at 800 °C TEM micrographs. (a) bright field; (b) TiC dark-field; (c) weak beam dark-field; and (d) fifth power of particle size versus time for an Fe–V–C alloy.[30] Courtesy of *Phil. Mag.*

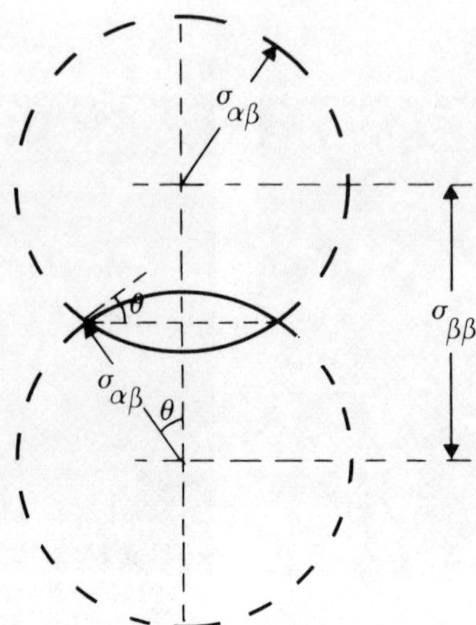

Fig. 14.   Assumed nucleus shape for precipitation on a planar grain boundary.[31]

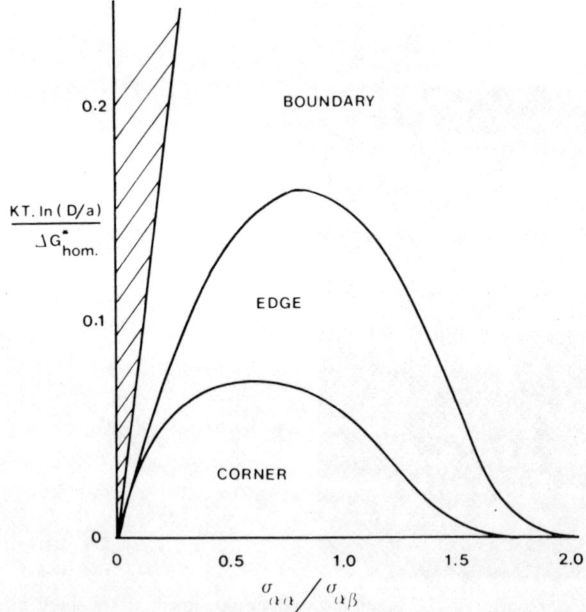

Fig. 15.   Conditions giving the greatest volume nucleation rates for various kinds of nucleation. The hatched area indicates where homogeneous nucleation is dominant.[32]

3.3 Planar Defects

We now consider the role of planar defects, e.g. grain boundaries, which are historically the most well known nucleation catalysts, although much of the theory may be applied to interphase interfaces.

The reduction in free energy obtained by nucleus formation at a grain boundary is in the ΔG_s term which is modified by subtracting the product of the grain boundary energy ($\sigma_{\alpha\alpha}$) and the area of boundary destroyed. The simple analysis due to Clemm and Fisher,[31] assumes a planar, disordered boundary and a nucleus shape as shown in Fig. 14, where the angle θ is governed by the relative values of $\sigma_{\alpha\beta}$ and $\sigma_{\alpha\alpha}$.

$$\Delta G = \Delta G_v \frac{2 \cdot \pi \cdot r^3}{3} (2 - 3\cos\theta + \cos^3\theta)$$

$$+ \sigma_{\alpha\beta} \cdot 4\pi r^2 \cdot (1 - \cos\theta) - \sigma_{\alpha\alpha} \cdot \pi r^2 \cdot \sin^2\theta \tag{20}$$

$r*$ is the same as for homogeneous nucleation and by assuming isotropic surface energy, so that $\cos\theta = \sigma_{\alpha\alpha}/2\sigma_{\alpha\beta}$,

$$\Delta G^*_{\text{het}} = \Delta G^*_{\text{hom}} \cdot 2f(\theta) \tag{21}$$

where $f(\theta) = \frac{1}{4}(2 - 3\cos\theta + \cos^3\theta)$ and is of the same form as $f(\alpha)$ in Fig. 6.

Thus the decrease in nucleation barrier is a function of θ, the nucleus–grain boundary contact angle.

Cahn[32] has extended this treatment to include grain edges (where three grains meet) and grain corners (where four grains meet), again assuming incoherent nuclei and isotropic surface energy. Figure 15 shows the conditions under which the highest volume nucleation rate for each kind of defect may be obtained, compared with the conditions for homogeneous nucleation. The ordinate represents the opposing factors; D/a, a ratio of grain diameter to grain boundary thickness, and the homogeneous nucleation barrier. The abscissa represents the catalytic effect of the defect (with $\sigma_{\alpha\alpha}/\sigma_{\alpha\beta} = 0$, denoting zero catalytic effect ($\theta = \pi$) and $\sigma_{\alpha\alpha}/\sigma_{\alpha\beta} = 2$, giving $\theta = 0$) and the zero nucleation barrier. (For example, for high driving forces (ΔG_v large) and low ratios of $\sigma_{\alpha\alpha}/\sigma_{\alpha\beta}$, homogeneous nucleation dominates, whilst corners may only dominate at very low driving forces.) The form of this diagram, of course, is a reflection of the ratios of the numbers of the potential nucleation sites. (For a grain size of 0·1 mm, for example, the ratio of lattice sites to grain corner sites is of the order of 10^{18}.)[33] Nucleation on grain boundaries is thus favoured under conditions of small ΔG_v together with a large ratio of grain boundary to precipitate–matrix interfacial energy.

Further reductions in the total interfacial energy contribution to the nucleation barrier may be effected if the nucleus adopts a coherent or semicoherent interface with one of the adjoining grains. This is likely to occur if there is a similarity in crystal structure and lattice parameter between precipitate and matrix, such that the precipitate is able to form with an orientation relationship with respect to one or both of the adjoining grains, e.g. see reference 34. For facetted nuclei at grain boundaries the potential reductions in interfacial energy have been treated theoretically by Clough *et al.*[35] using the same approach as for the facetted nucleus forming homogeneously (Section 2).

The contact angle between precipitate nucleus and grain boundary, depends on the ratio of the grain boundary energy, $\sigma_{\alpha\alpha}$, to the precipitate–matrix interfacial energy, $\sigma_{\alpha\beta}$. For maximum nucleation potency, a high value of this ratio is required, and this may be achieved if $\sigma_{\alpha\beta}$ corresponds to a low energy (coherent, or semicoherent) interface. Thus it is suggested that if the grain boundary plane lies parallel to a low energy precipitate–matrix interface, then nucleation will be especially favoured. The theoretical results of Clough *et al.*[35] strongly support this suggestion. As the angle between the precipitate habit plane and the grain boundary plane is reduced, there is a sharp decrease in activation barrier to nucleation. Calculations show that with a ratio of $\sigma_{\alpha\alpha}/\sigma_{\alpha\beta}$ of 5/3, the nucleation barrier disappears altogether when this angle becomes less than $\sim 26°$.

Increases in $\sigma_{\alpha\alpha}$, also serve to increase the ratio of $\sigma_{\alpha\alpha}/\sigma_{\alpha\beta}$ and hence favour nucleus formation. Theory therefore predicts that grain boundaries of high energy will be especially favoured.

3.3.1 Distribution of Grain Boundary Precipitates

It is well established, experimentally, that the number, distribution and morphology of grain boundary precipitates is widely variable from one grain boundary to the next, and now we consider the role of the boundary energy and structure.

The energy of a planar grain boundary varies with the misorientation of the adjoining grains in a manner analogous to that described for coherent, semicoherent and incoherent interphase interfaces. The following types may be classified:

(*i*) *Low angle boundaries* in which the misorientation is recognisably accommodated by dislocation networks with Burgers vectors characteristic of lattice dislocations. (Compare with semicoherent interphase interfaces.)

These boundaries are of low energy, and the major component of the grain boundary energy arises from the presence of the strain fields of the individual dislocations. With increasing misorientation, the spacings of the dislocations must decrease, and thus the energy increases. Eventually, the dislocation spacing becomes small enough for interference of the cores of neighbouring dislocations to occur, and this simple model of grain boundary structure is no longer tenable. In cubic lattices this situation occurs at a misorientation of 10–15°.

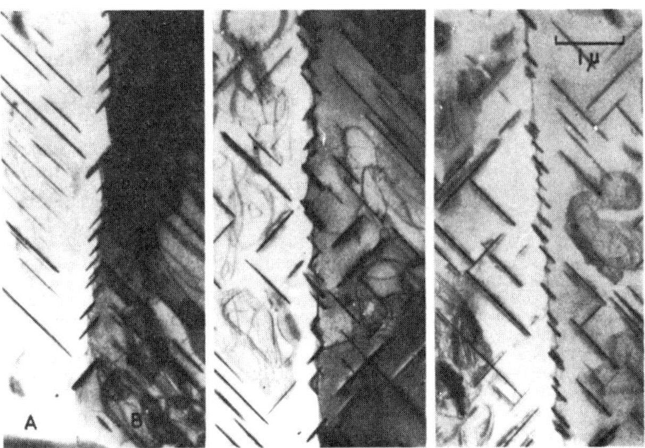

FIG. 16.   Precipitation of θ' at a curved, low angle boundary in an Al–Cu alloy. The boundary precipitates are related to grain A in the left-hand micrograph, those in the right-hand micrograph to grain B while the central micrograph contains boundary precipitates in both orientations.[36] Courtesy of *Acta Met.*

The chief consequence of this is that precipitate nucleation on low angle boundaries relies on the interaction between the nucleus and individual dislocations. Thus the nucleation process is controlled by changes in ΔG_{Strain} rather than $\Delta G_{Surface}$, and the nature, morphology and distribution of the precipitates may be interpreted in a similar manner to that described for individual matrix dislocations (see Section 3.2). For example, the nucleation potency is likely to be greatest for precipitates with coherent or semicoherent interfaces (i.e. transition precipitates) and appreciable strain energy components of nucleation. This has been illustrated by Vaughan[36] in a study of the Al–Cu system. At boundaries with misorientations of $<9°$, the θ' phase is nucleated (Fig. 16), rather than the equilibrium θ phase which forms at boundaries with misorientations $>9°$. Within the low angle boundary range there is also evidence that the nucleation rate increases with

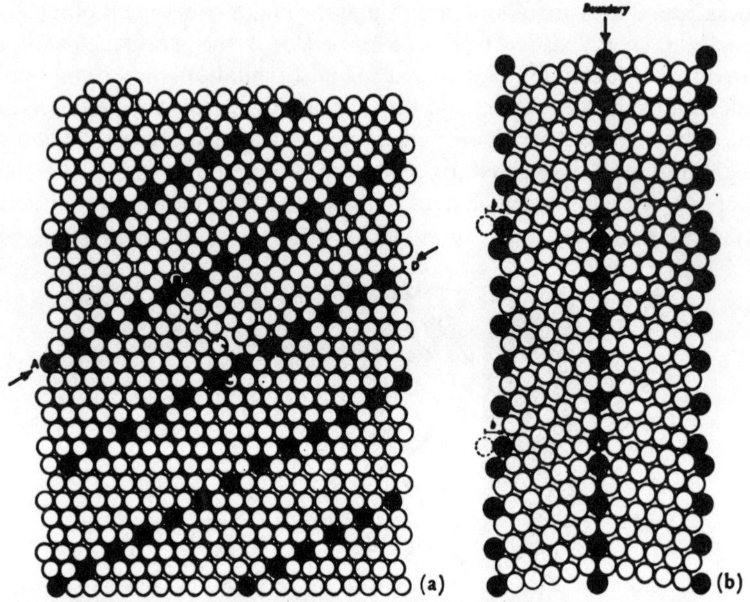

FIG. 17.   Diagrams showing two types of defects in a high angle coincidence boundary in a bcc bicrystal with a coincidence site density of 1 in 11: (a) Step where the boundary jumps from one close-packed plane in the coincidence site lattice to the next; (b) dislocations introduced to make the misorientation deviate slightly from perfect coincidence.[38] Courtesy of *Acta Met.*

misorientation, due to the increasing dislocation density in the boundary. Clark[37] has noted similar effects in the Al–Ag system, where the precipitate morphology is affected by the misorientation. Plates of the γ-phase were observed at boundaries with misorientations of $<16°$, with allotriomorphs being present at high angle boundaries.

(*ii*) *High angle boundaries* are generally characterised by high angular misorientations and random boundary planes. For a given misorientation, changes in boundary curvature vary the proportion of regions of good fit (where there is a high density of atoms at periodic sites) to regions of bad fit (where there is a high density of grain boundary ledges, steps and lattice dislocations). The grain boundary energy is thus a function of the boundary topography.

A further classification is possible, depending on whether the misorientation is completely random, or, corresponds to a *special* value, allowing a substantial measure of lattice matching to be achieved at the boundary. These latter boundaries are referred to as coincidence boundaries, and have lower energies than random boundaries. (A striking example is the fcc coherent twin boundary, across which the lattice matching is perfect; only the stacking sequence is changed.) The degree of coincidence is a measure of the proportion of lattice sites in the boundary which are common to the lattices of each of the adjoining grains. Small deviations from special misorientations may be accommodated by networks of dislocations (in a similar manner to low angle grain boundaries). See Fig. 17. These special boundaries have been shown to have lower diffusivities, higher mobilities, and a lower impurity segregation level than random grain boundaries.[39] (Since impurity segregation reduces the grain boundary energy, and occurs preferentially to random boundaries, the difference in grain boundary energy between special and random boundaries is probably only apparent in very high purity materials.)

We now attempt to interpret observations of grain boundary precipitate distribution in terms of the variation of both boundary misorientation and topography.

For low energy coincidence boundaries the theory predicts that the nucleation rate should be lower than for a random high energy boundary for the following reasons.

(i) The value of $\sigma_{\alpha\alpha}/\sigma_{\alpha\beta}$ should be smaller since $\sigma_{\alpha\alpha}$ is lower (unless solute segregation has occurred and thus lowered $\sigma_{\alpha\alpha}$ at random boundaries).

(ii) The lower segregation level and solute diffusivity in coincidence boundaries will reduce the rate at which atoms are able to join the critical nucleus.

The experimental evidence is rather conflicting, however. LeCoze *et al.*[40,41] have observed *higher* nucleation rates at certain coincidence boundaries in Al–Cu bicrystals, whereas Unwin and Nicholson,[42] and Butler and Swann[43] have noted *lower* nucleation rates at boundaries close to coincidence orientations in Al–Zn–Mg alloys. Butler and Swann have concluded that the lower segregation levels and solute diffusivities at coincidence boundaries are the most important factors.

The influence of boundary topography on precipitate nucleation rate is also well illustrated by the observations of Butler and Swann.[43] At curved, random, boundaries both precipitate number density and morphology were observed to vary with change in grain boundary plane, and precipitate nucleation was found to be especially favoured when there is a good

P. A. BEAVEN

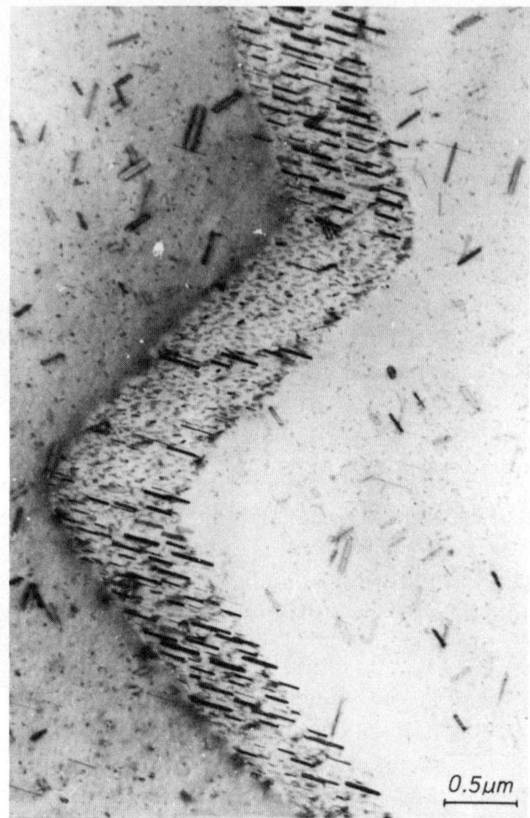

FIG. 18. TEM micrograph showing the influence of boundary plane inclination on preferred variant of η precipitate nucleated in Al–Zn–Mg alloy. Courtesy of E. P. Butler.

crystallographic matching between the precipitate habit plane and grain boundary plane (Fig. 18). Since the η phase in Al–Zn–Mg alloys may possess up to eight different orientation relationships with the matrix, good crystallographic matching between boundary and precipitate is relatively easy to achieve, thus accounting for the copious grain boundary nucleation observed. The results in the Al–Zn–Mg alloy are thus in agreement with the theoretical analysis of Clough et al.[35]

A further factor which affects the distribution of grain boundary precipitates is the presence of steps, ledges and lattice dislocations. Butler and Swann[43] have noted the absence of precipitation on some ledges (Fig.

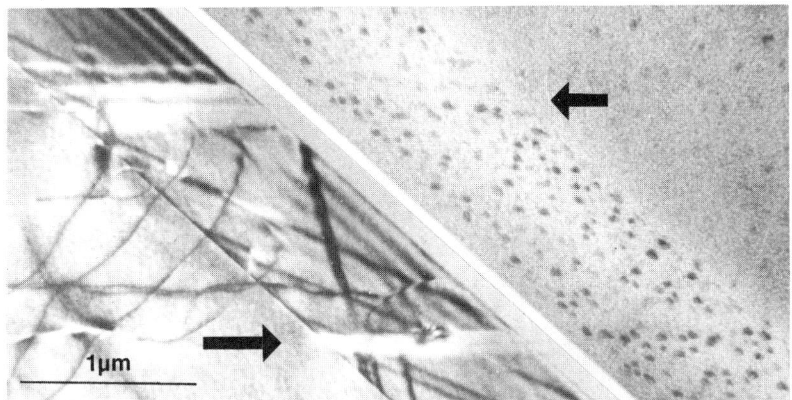

Fig. 19. TEM micrographs showing the absence of precipitation on grain boundary ledges but preferred nucleation on the edges of the ledges in Al–Zn–Mg alloy.[43] Courtesy of *Acta Met.*

19) and this is interpreted on the basis of the abrupt local change in grain boundary plane. Preferential nucleation along the edges of ledges was observed, however, and this may result from the presence of a strain field at the ledge, or from the ability of the nucleus to draw solute from the adjacent precipitate-free region. That lattice dislocations situated in grain boundaries may act as precipitate nucleation sites is best illustrated by the work of Jones *et al.*,[44] who have studied the effect of prior cold work on the precipitation of niobium carbide at grain boundaries in a commercial austenitic stainless steel. The distribution of niobium carbides at grain boundaries was shown to be dependent on the degree of cold work through the density of lattice dislocations present at the grain boundaries. This observation may be interpreted in terms of the high strain energy component of nucleation for niobium carbide (Section 3.2).

3.3.2 Morphology of Grain Boundary Precipitates

The examples referred to in the previous section have been mainly concerned with the nucleation of grain boundary precipitates where the crystallography of the matrix–precipitate orientation relationship plays an important role in determining the morphology of the precipitate. In this situation the effect of the grain boundary itself on precipitate morphology is often minor, although the influence on the spatial and size distributions of the resulting precipitate dispersion may be appreciable.

For typical, incoherent, equilibrium phases nucleated at grain boundaries,

                    P. A. BEAVEN

the observed morphology often approaches that assumed in the simple theoretical models. Nucleation is thought to occur by either of the two mechanisms depicted schematically in Fig. 20. In the first case, nucleation occurs on a planar section of grain boundary which remains stationary during the process. The activation energy for nucleation is, of course, a strong function of nucleus shape and recent calculations suggest that a 'pill-box' morphology is the most appropriate.[45] The characteristic

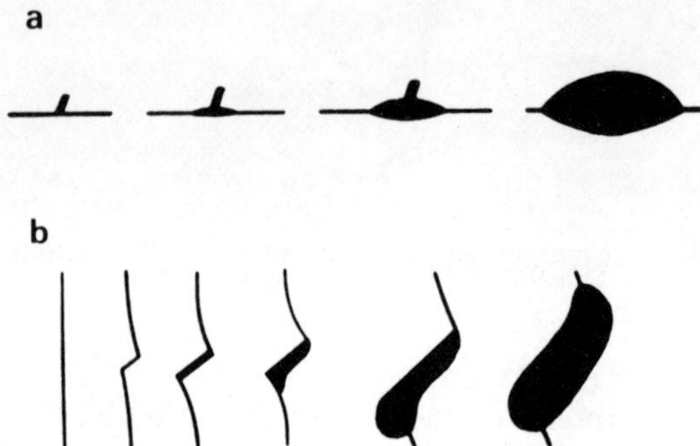

FIG. 20.   Schematic diagrams showing; (a) the conventional, and (b) the 'pucker', mechanisms for nucleation at grain boundaries.

allotriomorph shape is then developed during subsequent growth with incoherent interfaces being formed with respect to both of the adjoining grains. In the second case, it is envisaged that the grain boundary is displaced so that a broad face of the nucleus may abut the boundary—the so-called 'pucker mechanism'.[46] In order for this to occur the saving in interfacial energy must be sufficient to offset the increase in grain boundary energy resulting from the increase in grain boundary area. A consequence of this is that the precipitate possesses a low energy interface with respect to one of the grains, and generally, growth may occur more rapidly by the movement of the incoherent interface into the other grain. A detailed comparison of these models has been made by Aaronson and Aaron[47] in connection with the initiation of cellular precipitation.

As already mentioned, further reductions in interfacial energy may be achieved if the precipitate develops low energy facets with one or both of the adjoining grains. The morphology thus depends not only on the crystallographic relationship between precipitate and matrix, but also on the orientation relationship between the adjacent grains. Figure 21 shows an example of the various morphologies which may develop when the bcc α

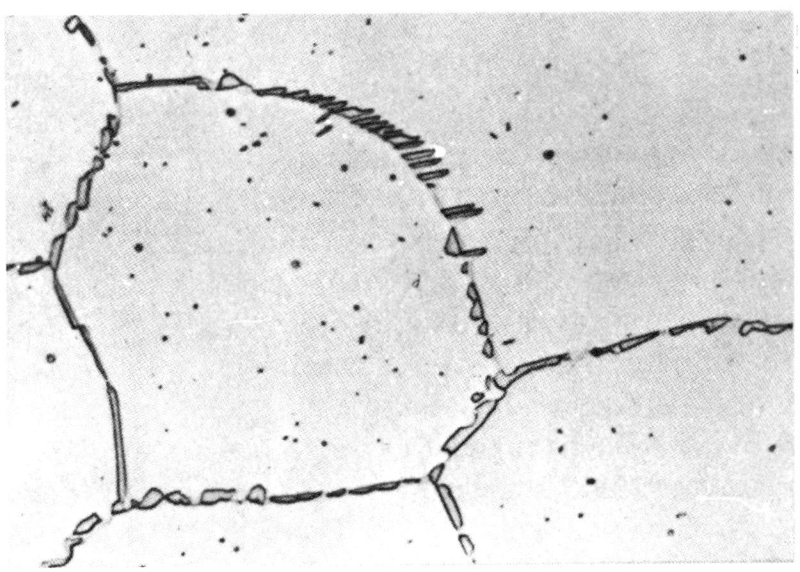

FIG. 21. Grain boundary precipitate morphologies in a Co–Fe alloy.[34] Courtesy of *Acta Met.*

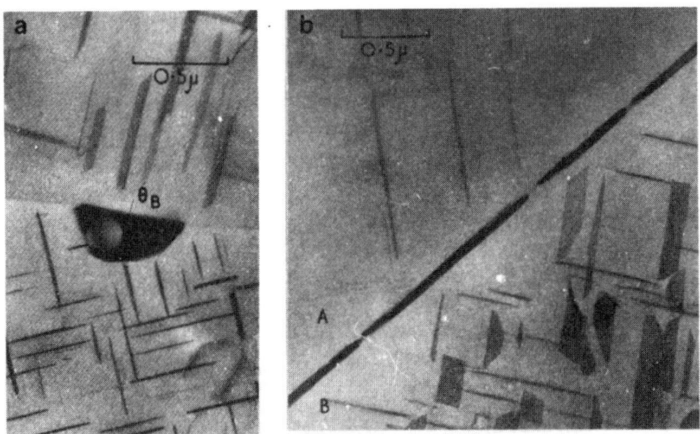

FIG. 22. Effect of boundary misorientation on θ morphology. In (a) the θ precipitate is orientation related to the upper grain and growth occurs into the lower grain. In (b) the particles have orientation relationships with both grains.[36] Courtesy of *Acta Met.*

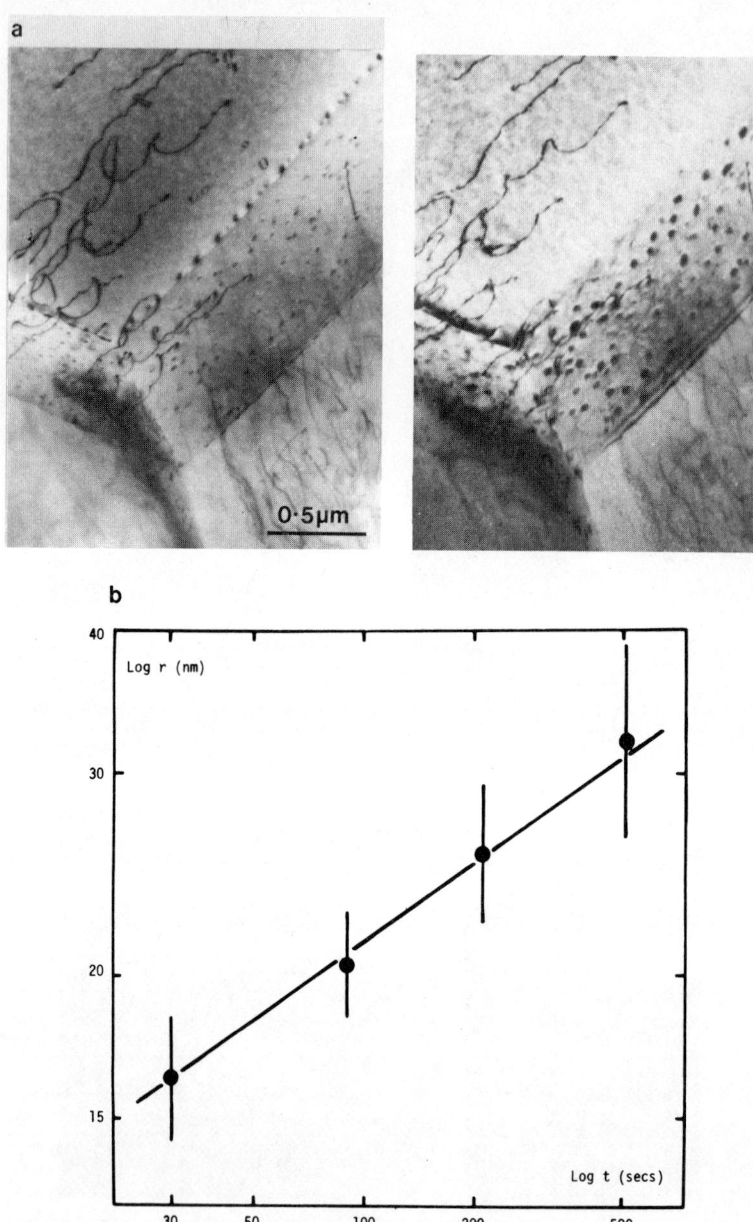

FIG. 23.   (a) Sequence of micrographs showing coarsening of particles on a high angle grain boundary in an Al–Zn–Mg alloy. (b) Log–log plot of particle radius versus time, with $n = 4\cdot4$.[43] Courtesy of *Acta Met*.

phase forms at grain boundaries of the fcc γ phase in a Co–Fe alloy.[34] Generally, the α phase is orientation-related to the grain in which nucleation has occurred, and has a random orientation with respect to the other grain into which growth takes place. Occasionally, however, the relative orientations of the two grains may be such as to allow the precipitate to be orientation-related to both grains. This situation is shown in Fig. 22, for the equilibrium θ phase at grain boundaries in Al–Cu, and illustrates the resulting precipitate morphologies.

The precipitate distributions and morphologies which may occur at different grain boundaries are thus seen to arise from a number of factors, and it is the wide variety of behaviour which hinders both experimental study and theoretical modelling.

3.3.3 Effect of Grain Boundaries on Precipitation Kinetics

The presence of high diffusivity paths along grain boundaries, together with possible prior segregation of solute atoms, may increase rates of nucleation, as has been considered already for dislocations. That allotriomorph growth rates may be higher than predicted from volume diffusion considerations has been demonstrated by Aaron and Aaronson[48] for the θ phase in Al–Cu. A model for the growth kinetics of grain boundary allotriomorphs has been formulated by Brailsford and Aaron,[49] in which direct volume diffusion to the allotriomorph determines the growth rate at high homologous temperatures, whereas at low temperatures (where $D_{GB} \gg D_v$) the growth mechanism involves solute diffusion along the grain boundary and interphase diffusion over the surface of the allotriomorph. Experimental investigations[48,50] have demonstrated the circumstances under which volume or grain boundary diffusion may predominate.

A theoretical treatment of the coarsening of grain boundary precipitates, assuming grain boundary diffusion[51] indicates particle size dependencies on $t^{1/4}$ for high angle boundaries, and a transition form $t^{1/5}$ to $t^{1/4}$ kinetics for low angle dislocation boundaries. The experimental investigations of Butler and Swann[43,52] using Al–Zn–Mg alloys and the *in situ* electron microscopy technique have indicated coarsening exponents between $n = 3$ and 4 at different grain boundaries. It is therefore suggested that the coarsening process occurs by cooperative diffusion through the matrix and along grain boundaries under the conditions of these experiments. Variations in the rate constant for coarsening have been interpreted in terms of the dependence of the grain boundary diffusivity on the angular misorientation between the grains.[52] A typical sequence of micrographs is shown in Fig. 23; the coarsening exponent has a value of 4·4 in this example.

4. CLOSURE

The role of lattice defects in modifying precipitation has been described in terms of the parameters controlling nucleation. The chief obstacle to the quantitative prediction of nucleation rates is the lack of information about these parameters, in particular surface energy terms. The size and spatial distributions, and morphology, of precipitates depends chiefly on the thermomechanical history (which controls the defect type, density, and distribution), and the competition between nucleation sites. Heterogeneous or homogeneous nucleation may dominate, depending on the interplay between both thermodynamic and kinetic factors. The resulting micro-structural inhomogeneities with respect to the size, nature, and spatial distribution of precipitates formed must clearly be taken into account when considering long term microstructural stability.

5. ACKNOWLEDGEMENTS

Acknowledgement is made to colleagues for the provision of micrographs, and to the editors of *Acta Metallurgica* and *Philosophical Magazine* for permission to reproduce figures.

REFERENCES

1. *Nucleation*, (1969). (Ed. A. C. Zettlemoyer), Marcel Dekker Inc., NY.
2. Eshelby, J. D. (1957). *Proc. Roy. Soc.*, **A241**, 376.
3. Brown, L. M., Woolhouse, G. R. and Valdre, U. (1968). *Phil. Mag.*, **18**, 781.
4. Russell, K. C. (1970). In *Phase Transformations*, ASM Metals Park, Ohio, p. 219.
5. Nabarro, F. R. N. (1940). *Proc. Roy. Soc.*, **175**, 519; (1940). *Proc. Phys. Soc.*, **52**, 90.
6. Servi, I. S. and Turnbull, D. (1966). *Acta Met.*, **14**, 161.
7. Kirkwood, D. H. (1970). *Acta Met.*, **18**, 563.
8. Nicholson, R. B. (1971). In *Strengthening Mechanisms in Crystals*, (Ed. A. Kelly and R. B. Nicholson), Applied Science Pub. Ltd., London.
9. Lee, J. K. and Aaronson, H. I. (1974). *Scripta Met.*, **8**, 1451.
10. Nicholson, R. B. (1974). *Scripta Met.*, **8**, 269.
11. Shepherd, J. P. (1976). *Metal Science*, **10**.
12. Herman, H. (1971). *Met. Trans.*, **2**, 13.
13. Cahn, J. W. (1957). *Acta Met.*, **5**, 169.
14. Katgerman, L. and Van Liere, J. (1978). *Acta Met.*, **26**, 361.
15. Gomez-Ramirez, R. and Pound, G. M. (1973). *Met. Trans.*, **4**, 1563.
16. Lyubov, B. Ya. and Solovyev, V. A. (1965). *Phys. Met. Metallog.*, **19**, 13.
17. Dollins, C. C. (1970). *Acta Met.*, **18**, 1209.

18. Russell, K. C. and Aaronson, H. I. (1975). *J. Mat. Sci.*, **10**, 1991.
19. Aaron, H. B. and Aaronson, H. I. (1971). *Met. Trans.*, **2**, 23.
20. Beaven, P. A. and Butler, E. P. (1976). *Proc. 34th E.M.S.A. Annual Meeting*, Claitor's Pub. Div., Los Angeles, USA.
21. Johari, O. and Thomas, G. (1965). *Trans. ASM*, **58**, 563.
22. Rowcliffe, A. F. and Nicholson, R. B. (1972). *Acta Met.*, **20**, 43.
23. Beaven, P. A. and Polmear, I. J. to be published.
24. Headley, T. J. and Hren, J. J. (1976). *Phil. Mag.*, **34**, 101.
25. Chaturvedi, M. C., Honeycombe, R. W. K. and Warrington, D. H. (1968). JISI, **206**, 1236.
26. Beaven, P. A., Swann, P. R. and West, D. R. F. (1978). *J. Mat. Sci.*, **13**, 691.
27. Raty, R. and Miekk-oja, H. M. (1968). *Phil. Mag.*, **18**, 1105.
28. Beaven, P. A. and Butler, E. P. (1979). 'Phase Transformations', Proc. IOM. Conf., York.
29. Kreye, H. (1970). *Z. Metall.*, **61**, 108.
30. Dunlop, G. L. and Honeycombe, R. W. K. (1975). *Phil. Mag.*, **32**, 61.
31. Clemm, P. J. and Fisher, J. C. (1955). *Acta Met.*, **3**, 70.
32. Cahn, J. W. (1956). *Acta Met.*, **4**, 449.
33. Christian, J. W. (1975). *The Theory of Transformations in Metals and Alloys*, Part I, 2nd edn, Pergamon Press, Oxford.
34. Ryder, P. L., Pitsch, W. and Mehl, R. F. (1967). *Acta Met.*, **15**, 1431.
35. Clough, S. P., Lee, J. K., Bradley, J. R., Carlson, J. A., Large, W. F., Micheals, K. F., Seaton, C., Aaronson, H. I. and Russell, K. C. (1974). *Scripta Met.*, **8**, 791.
36. Vaughan, D. (1968). *Acta Met.*, **16**, 563.
37. Clark, J. B. (1965). In *High-Temperature, High Resolution Metallography*, (Ed. H. I. Aaronson and G. S. Ansell), Gordon and Breach, New York.
38. Brandon, D. G. (1966). *Acta Met.*, **14**, 1479.
39. Pumphrey, P. H. (1976). In *Grain Boundary Structure and Properties*, (Ed. G. A. Chadwick and D. A. Smith), Academic Press, London.
40. LeCoze, J., Biscondi, M., Levy, J. and Goux, C. (1973). *Mem. Scient. Revue Metall.*, **70**, 397.
41. LeCoze, J. and Biscondi, M. (1974). *Can. Met. Q.*, **13**, 59.
42. Unwin, P. N. T. and Nicholson, R. B. (1969). *Acta Met.*, **17**, 1379.
43. Butler, E. P. and Swann, P. R. (1976). *Acta Met.*, **24**, 343.
44. Jones, A. R., Howell, P. R., Page, T. F. and Ralph, B. (1974). Grain Boundaries in Engineering Materials, Fourth Bolton Landing Conference, Lake George, New York.
45. Aaronson, H. I. (1979). 'Phase Transformations', Proc. IOM Conf., York.
46. Tu, K. N. and Turbull, D. (1967). *Acta Met.*, **15**, 1317.
47. Aaronson, H. I. and Aaron, H. B. (1972). *Met. Trans.*, **3**, 2743.
48. Aaron, H. B. and Aaronson, H. I. (1968). *Acta Met.*, **16**, 789.
49. Brailsford, A. D. and Aaron, H. B. (1968). *JAP*, **40**, 1702.
50. Hawbolt, B. and Brown, L. C. (1967). *Trans. Met. Soc. AIME.*, **239**, 1916.
51. Ardell, A. J. (1972). *Acta Met.*, **20**, 601.
52. Butler, E. P. and Swann, P. R. (1975). In *Physical Aspects of Electron Microscopy and Microbeam Analysis*, (Eds. B. Siegel and D. R. Beaman), Wiley, New York, p. 129.

4

Influence of Service Conditions on Phase Development and Transformation*

G. SAUTHOFF

*Max Planck Institut für Eisenforschung GmbH,
Düsseldorf, FRG*

1. INTRODUCTION

The structures of materials used in practice are unstable with respect to thermodynamics. Without special external forces, grain boundaries and dislocations are not equilibrium crystal defects, and likewise the distributions of phases do not correspond to equilibrium. Consequently, in reality, the structures are continually changing, a fact which has been investigated for a multitude of cases.[1] The structures strive to reach the equilibrium state with minimum total grain boundary area and dislocation length, without metastable phases and with the stable phases coagulated into single large particles, so that the total (Gibbs) energy attains its minimum value. The rates of these structure changes are appreciable at elevated temperatures, especially under conditions of hot deformation or creep; they are influenced by other physical variables, besides temperature, which allow the adjustment to specific requirements within certain limits. On this background 'structure stability' has only a relative sense; a structure is called stable if the structure changes are sufficiently slow with respect to the applications.

The physics of the structure changes were discussed recently in much detail.[2] Here the important points concerning hot deformation and creep will be noted with regard to conventional materials, especially from the viewpoint of the effect of concurrent deformation. Furthermore, conclusions will be drawn and criteria will be given for deciding the best

* Revised translation of 'Stabilität und Optimierung des Gefüges' in *Festigkeits- und Bruchverhalten bei höheren Temperaturen* (Ed. W. Dahl and W. Pitsch), Verlag Stahleisen GmbH, Düsseldorf, 1980, vol. 1, 100–148.

93

action to take to optimise stable structures and for estimating the order of magnitude of the effects.

2. STABILITY OF GRAIN STRUCTURE

The grain size of a polycrystalline alloy is of different significance for different hot deformation or creep mechanisms; this is illustrated by the grain size dependence of the 'deformation maps' (stress–temperature–deformation diagrams).[3] At elevated temperatures ($T/T_M > 0.3$, $T_M =$ temperature of melting) the grain structure of single-phase alloys coarsens,[4] i.e. the average grain size increases and the grain number decreases with time so that under conditions of hot deformation or creep the grain structure is unstable. Consequently the deformation behaviour can be shown directly by grain size–stress–deformation diagrams.[5]

A simultaneous deformation during grain coarsening does not affect the grain structure as long as it only actuates recovery. If, however, the deformation rate is in a critical range of comparatively high creep rates, dynamic recrystallisation can be initiated[3] so that the existing grain structure is replaced continually by a new one. Dynamic recrystallisation has been discussed in detail repeatedly, see, for example, reference 6.

Precipitated particles of a second phase impede the movement of grain boundaries so that a grain structure can be stabilised by a suitable particle dispersion.[4,7] A detailed theoretical discussion[8] leads to the following criterion for grain stability: a grain structure which is characterised by a grain diameter d, can be stabilised by particles of diameter $2r$ if

$$\frac{d}{2r} \leq \frac{\alpha}{f} \tag{1}$$

where f is the precipitated volume fraction and α a factor of the order of 0.5. (This factor is not completely understood. It depends on d, r and f; in particular it decreases with f for $f \geq 10\%$.) The stabilised grain size is then given by eqn (1) with an equal sign. Experiments with a plain carbon steel[8] or a complex duplex structure[9] confirm the criterion.

However, the grain stability as produced by a particle dispersion is not absolute. It is possible that particles are dragged by moving boundaries like segregated atmospheres[10,11] where diffusion is rate controlling. Here, however, this effect is not of practical importance particularly as there is self-stabilisation: the driving force decreases with grain coarsening, and the number of dragged particles which are 'picked up' by the moving boundary is increased. The coarsening of the stabilising particles which is discussed in Section 3 cannot be neglected and leads to a slow grain coarsening according to eqn (1). Apart from this the danger of abnormal growth[4,8] has to be taken into account: if a grain is substantially larger than the surrounding grains—especially if it is larger than the stabilised grain size— then it grows into the surrounding grain structure. This instability can only be avoided if the stabilising particle dispersion contains a sufficient number of larger particles.

3.   PHASE STRUCTURE STABILITY

3.1 A Small Amount of One Precipitate Phase

In reference 12 the fundamentals of the theoretical description of precipitation processes and their kinetics are discussed. Some significant points from this work concerning the stability of precipitates will be noted in the following. The following assumptions are made initially: Spherical particles of a pure component or compound with radii r are precipitated, the total precipitated volume fraction is small so that the particle distances are always large compared with r, and the matrix behaves as a very diluted solution. It was shown[12] that the particle solubility—and therefore the growth rate—changes during precipitation because of its size dependence and that the particles do not grow isolated, but dependent on one another in mutual competition.

Figure 1 shows schematically the local concentration c as a function of the distance variable ρ between two growing particles with radii r_1 and r_2 at

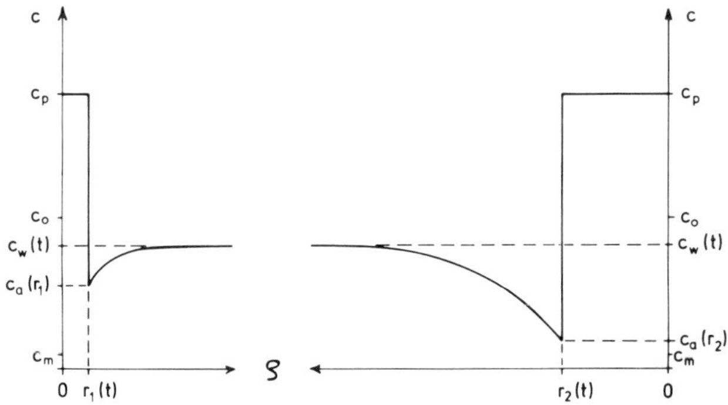

FIG. 1.   Concentration c as a function of distance ρ for two growing particles (schematic, see text).[12]

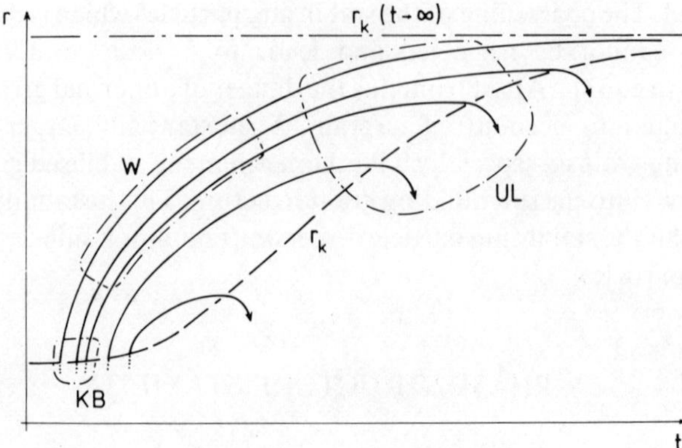

FIG. 2. Time dependence of the particle radii $r(t)$ (schematic, see text).[12]

time t (c_p is the concentration within the particles, c_0 is the initial concentration of the supersaturated matrix, c_w is the instantaneous concentration in the matrix, c_a is the particular concentration in equilibrium with the particle, i.e. the particle solubility, and c_m is the saturation concentration of the matrix). Because of its lower solubility the larger particle grows faster than the smaller one and therefore at the expense of the smaller one.

The time dependence of the sizes of the growing particles is schematically shown in Fig. 2 by the 'growth paths' $r(t)$ (solid lines). In addition Fig. 2 shows the behaviour of the critical radius r_k (broken line) which characterises the instantaneous supersaturation c_w/c_m of the matrix and is given by

$$r_k = 2\gamma \left(c_p R T_A \ln \frac{c_w}{c_m} \right)^{-1} \tag{2}$$

(γ = interface energy per unit area, R = gas constant, T_A = precipitation temperature). After nucleation (domain KB in Fig. 2) the supersaturation is diminished by the precipitation, especially during the growth stage (domain W), so that r_k increases. Correspondingly the growth rates of all particles decrease after nucleation. Because of the size dependence of the particle solubilities the oldest particles grow with maximum velocity and remain the biggest ones whereas the particles with $r < r_k$ dissolve. After the exhaustion of the matrix the particle dispersion coarsens (domain UL of Ostwald ripening), i.e. the bigger particles grow at the expense of the smaller ones so that the average particle size increases, the particle number decreases, and the total interface energy is reduced. Then the average centre-to-centre distance ζ between the particles in the volume is

$$\zeta = 2\bar{r}f^{-1/3} \tag{3}$$

in simple approximation, and the particle number N per unit volume is

$$N = f \cdot (\tfrac{4}{3}\pi\bar{r}^3)^{-1} \tag{4}$$

(f = volume fraction to be precipitated, $\bar{r}$ = average particle radius).

Equation (3) can be used as a criterion for roughly discriminating between the precipitation stages. If

$$\zeta \gg 2\bar{r}f^{-1/3}, \tag{5}$$

then the crystal is still supersaturated, and further nucleation processes can be expected. If on the other hand

$$\zeta \simeq 2\bar{r}f^{-1/3}, \tag{6}$$

then nucleation is no longer possible, and the precipitate coarsens. According to the laws for Ostwald ripening[12] the time $\hat{t}$ for increasing the average particle diameter by a factor m—and decreasing the particle number by m^3—is

$$\hat{t} = (m^n - 1)\tau r_0^n \tag{7}$$

where r_0 is the initial radius, and the coarsening constant τ and the exponent n depend on the specific coarsening mechanism.[12] As long as $\hat{t}$ is longer than the service time $\hat{t}_e$—or m is smaller than the tolerable coarsening ratio m_t— then the structure can be regarded as stable. Accordingly, the stability criterion is

$$(m_t^n - 1)\tau r_0^n \geq \hat{t}_e \tag{8}$$

which can be used as long as the particle number is sufficiently large,[12] i.e. ζ (eqn (6)) is small compared with the specimen size. (If ζ is in the range of the specimen size, there would be a final state with a single particle of constant size $r = r_k(t \to \infty)$ in stable equilibrium with the mixed crystal. This is never attained in practice.)

If in the assumed simple case the diffusion between the particles determine the precipitation rate, then

$$n = 3$$

and

$$\tau = \frac{9}{8}\frac{c_p^2 RT_A}{D\gamma c_m} \tag{9}$$

where D is the (effective) diffusion coefficient which determines the temperature dependence of τ (see Section 3.3). Thus, for a stable precipitate structure it is not only a small diffusion coefficient which is decisive, but also a high solute concentration within the particles, a small interface energy as with coherent particles and a low saturation concentration c_m of the matrix corresponding to a small particle solubility. Since in many alloys it is possible to adjust the composition so that $c_m \ll c_p$, the variable c_m is of particular practical importance.

Furthermore, eqn (8) indicates that the precipitate structure becomes less stable with decreasing particle size, r_0. In an analogous way this is valid also for the growth stage. In other words the early precipitation stages, especially the growth at the start of coarsening, are always fast in comparison to the later ones. This is reflected by the decreasing slope of the 'growth paths' in Fig. 2 and corresponds to experience.

Apart from this, the mutual influence of nucleation, growth and coarsening has to be considered with respect to the consequences for the precipitate structure. If—compared with growth and coarsening—the nucleation is fast because of high supersaturation, many small nuclei are formed and this according to eqns (3) and (4) leads to a fine particle dispersion with an early commencement of the slower coarsening process. If because of nucleation difficulties the nucleation rate is small compared with the growth rate, only a few big nuclei are formed which leads to a coarse particle dispersion with a late commencement of slower coarsening.

3.2 A Precipitate with a Misfit and Many Components

A misfit between particle and matrix generates coherency stresses which yield additional contributions to the particle energy and thereby influence the particle solubility. As discussed in reference 13, the particle solubility c_a, i.e. the solute concentration in the matrix at the particle surface in equilibrium with the particle of radius r, can generally be expressed as

$$RT_A c_p \ln \frac{c_a}{c_m} = \frac{2\gamma}{r} + \Sigma E_i \tag{10}$$

which can be regarded as the generalised Gibbs–Thomson(–Freundlich) equation. The energies E_i are the additional energies which result, for example, from the coherency stresses (elastic self-energy) or from the interaction with external forces, (e.g. the creep stress), and depend on particle shape and distribution.[14,15]

The misfit between a small particle and the matrix can be accommodated elastically as long as it is smaller than about 5%.[16] Accordingly the interface is coherent in such cases with an interface energy in the range of $20\,\mathrm{mJ/m^2}$.[12] With a larger misfit the interface contains misfit dislocations (Fig. 3) so that the interface is now semi-coherent with a higher energy. When the particles grow, the coherency decreases by the insertion of additional misfit dislocations which reduce the coherency stresses.[16] The

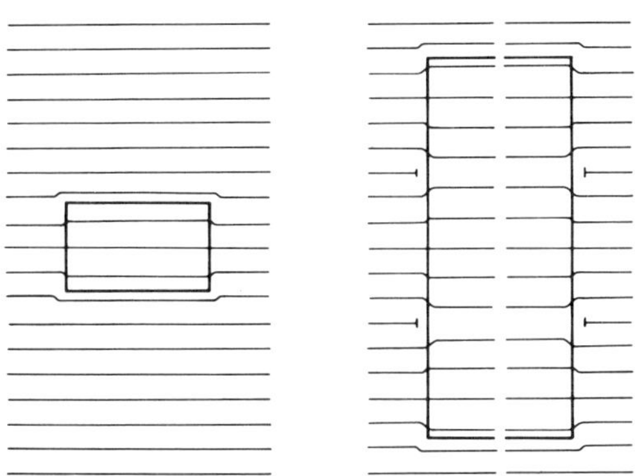

FIG. 3.   Coherent particle and semi-coherent particle with misfit dislocations (schematic).

misfit, δ, that is accommodated elastically and is attributable to the residual coherency stresses can be estimated by means of the Brooks criterion:[17] a dislocation is inserted into the interface when the absolute misfit becomes larger than the Burgers vector, $\mathbf{b}$, so that the order of magnitude is[13]

$$\delta = \frac{\mathbf{b}}{2r} \tag{11}$$

Therefore, since the dislocation density remains constant, the interface energy is not affected in this case.

With anisotropic coherency stresses and interface energies the optimum particle shape is not a sphere, but is for example a cube, disc or needle. The usual expressions, especially eqn (10), are then still valid, if r and γ are regarded as effective variables.[14]

So far the simple case of the precipitation of a pure component, or compound, in a binary alloy has been discussed so that the particle solubility is given by the saturation concentration of solute in the matrix, and the Gibbs–Thomson equation has the simple form of eqn (10). If the precipitated phase is not pure, but has an extended range of composition, the Gibbs–Thomson equation is more complex[18] and eqn (10) is valid only in an approximate way which is, however, sufficient in many practical cases.

In alloys and precipitates with more than two components the saturation concentration is replaced by the solubility product with a corresponding modification of the Gibbs–Thomson equation.[18,19] If, however, the ratio of the concentrations in the matrix is the same as in the precipitate (stoichiometric composition), the matrix composition can be characterised by one effective concentration variable, and eqn (10) is valid.[19] Likewise the situation is simple if, for example, in a ternary alloy, one component exists in such great abundance that its amount is hardly changed by precipitation. Then the particle solubility can be approximated by the saturation concentration of the minority component, i.e. the alloy can be treated as a binary one.[19]

As an example of the effect of the relative parameters on the precipitate stability, a detailed analysis of the coarsening of cementite in a low alloy steel[20] is noted which emphasises the strong dependence of the precipitation kinetics on the solubility conditions. It has already been observed in the preceding section that solubility is the most important variable for the stability control. Particularly in alloys with several components the coarsening rate can be varied to a large degree by varying the alloy composition, as was shown for the precipitation of SiO_2 in copper with various oxygen concentrations.[21]

3.3 Diffusion Coupling

In addition to the driving force—as determined by interface energies and particle solubilities—the kinetic factors, i.e. the diffusion coefficients and the rate constants of the reactions at the particle interfaces, also control the precipitate stability. To begin with it is assumed that the diffusion of solute in the matrix between the particles is the slowest step that determines the rate of precipitation or coarsening. Then for estimating the stability of the structure, e.g. by using eqn (8), the effective diffusion coefficient D has to be known. However, D is a complex variable even in binary alloys because of the coupling of the diffusion fluxes; this diffusion coupling will now be discussed.

At a growing particle, solute atoms diffuse from the matrix to the particle, and solvent atoms diffuse vice versa. If the diffusion fluxes were independent, the atoms would diffuse according to their specific concentration gradients and diffusion coefficients. The atom fluxes correspond to volume fluxes and would be dissimilar as long as the atomic volumes and the diffusion coefficients of both components were not identical. Thus independent partial fluxes would lead to differences in volume, i.e. to additional lattice distortions and corresponding coherency stresses. Actually the partial fluxes adjust themselves in an optimum way so that the additional coherency stresses do not become too large (see Kirkendall effect). Thus the stresses enforce the coupling of the diffusion fluxes which in turn determines the value of the effective diffusion coefficient.[22] The degree of coupling is diminished if the additional coherency stresses are relieved by secondary mechanisms, e.g. by the exchange of vacancies with neighbouring boundaries or climbing dislocations.

Diffusion coupling has been analysed for an isotropic AB alloy with spherical β particles in an α matrix as a function of the degree of coupling so that expressions for the effective diffusion coefficient and for the hydrostatic coherency pressure are available, as are expressions dealing with particular relaxation mechanisms.[22] If there is no relaxation (complete coupling), the coherency stress is at its maximum, and D in eqn (9) is expressed by

$$D = \hat{D} = \frac{D_A^\alpha D_B^\alpha (V_m^\alpha)^2}{x_A^\alpha D_A^\alpha V_A^2 + x_B^\alpha D_B^\alpha V_B^2} \tag{12}$$

with

$$V_A = \frac{x_B^\beta V_m^\alpha - x_B^\alpha V_m^\beta}{x_B^\beta - x_B^\alpha} \qquad V_B = \frac{x_A^\alpha V_m^\beta - x_A^\beta V_m^\alpha}{x_A^\alpha - x_A^\beta}$$

where D_A^α, D_B^α and V_m^α, V_m^β are the diffusion coefficients of the respective components and the volumes of the respective phases per atom (or per mole), and x is the respective atom (or mole) fraction. If on the contrary, there is complete relaxation (decoupling), the coherency stresses disappear, and D in eqn (9) is given by Darken's interdiffusion coefficient:

$$D = \tilde{D} = x_B^\alpha D_A^\alpha + x_A^\alpha D_\beta^\alpha \tag{13}$$

This discussion is of minor importance for substitutional alloys since in most practical cases the diffusion coefficients and the atomic volumes of the different components have about the same order of magnitude. Therefore the diffusion is independent of coupling and is controlled by the minority component:

$$D \approx \hat{D} \approx D \approx D_B^\alpha$$

The discussed relationships are important for interstitial alloys, e.g. steels (Fe–C) where D_{Fe} and D_C differ by several orders of magnitude. It follows from eqns (12) and (13) in this case that

$$D_C^\alpha \approx \tilde{D} \gg D \gg \hat{D} \gg D_{Fe}$$

With regard to the fast carbon diffusion, a stress relief by a still faster volume relaxation can not usually be expected so that $D \approx \hat{D}$ seems to be a suitable approximation (see below).

If in a ternary alloy with components A, B and C a compound Z is precipitated according to the chemical reaction $b\mathrm{B} + c\mathrm{C} \rightarrow z\mathrm{Z}$, the ratio of the fluxes of B and C atoms to the particle must be equal to b/c, i.e. there is an additional stoichiometric coupling leading to[18]

$$D_Z = \frac{(x_B^\alpha x_C^\alpha)^{1/2} D_B^\alpha D_C^\alpha (bc)^{1/2} z}{c^2 x_B^\alpha D_B^\alpha + b^2 x_C^\alpha D_C^\alpha} \tag{14}$$

In reference 18 the complete expression for τ, eqn (9) (with a different normalisation of the fluxes) can be found for an oxide (or carbide) of low solubility. This expression considers simultaneous stoichiometric and stress coupling.

The coherency stresses that result from the different diffusion fluxes ('coupling pressure'), influence not only the effective diffusion coefficient, but also the particle solubility because according to eqn (10) the particle solubility c_a increases with increasing elastic energy E_i of the particle. According to the cited analysis[22] the order of magnitude of the 'coupling pressure' can be estimated, especially when there is no relaxation (complete coupling). In substitutional alloys, with D_A and D_B in the same range of order the 'coupling pressure' is small compared with the 'precipitation pressure' $2\gamma/r_k$ (see eqn (2)) which determines the precipitation behaviour and the particle growth.[12] In such a case the solubility change is therefore negligible for the particle behaviour. If, however, $x_B^\alpha D_B^\alpha \gg x_A^\alpha D_A^\alpha$ as in interstitial alloys, then the 'coupling pressure' is in the same range of order as $2\gamma/r_k$. In this case the precipitation would stop if there were no stress relaxation; in other words the particles grow only at the rate at which the 'coupling pressure' is relaxed by secondary processes.

Under such circumstances the precipitation rate is no longer determined by the diffusion in the matrix, but by a slower interface reaction. It can generally be stated that the precipitation that is controlled by an interface reaction is always slower than a diffusion controlled precipitation. Thus the diffusion controlled rate is the upper limit for precipitation rates. It should be noted that the diffusion flux to the particle usually decreases with increasing particle size to a greater extent than does the reaction flux through the interface.[19] Consequently the interface reaction can only control the precipitation rate of sufficiently small particles depending on the ratio of the respective rate constants.

3.4 Several Precipitate Phases

In creep resistant alloys, e.g. superalloys or steels, several phases precipitate together as illustrated by the schematic in Fig. 4. The different solubilities of the different phases lead to different precipitation kinetics so that one phase may already be in the pure coarsening stage while another is still nucleated. The different solubilities of the phases mean different phase stabilities, i.e. a precipitation sequence results, as is well known, particularly for steels. The different precipitate phases that are mutually

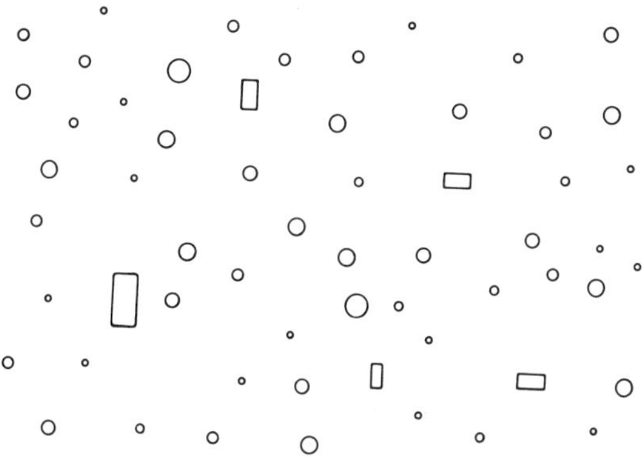

Fig. 4. Concurrent precipitation of two phases with ($\bigcirc$) particles and ($\square$) particles (schematic).

soluble, are 'coupled' by the common matrix so that they influence each other's precipitation behaviour. Phases with different solubilities can exist together only for a limited time, i.e. until the less stable phase has completely dissolved in favour of a more stable one. Thus, in addition to the exchange of solute atoms between particles of one phase which leads to particle coarsening,* there is an exchange of solute atoms between particles of different phases. This 'general Ostwald ripening' where the controlling solubility differences result from differences in size, stress state and phase, will be known as 'transcrystallisation' in the following. (The possibilities of describing the general precipitation case with many components and phases have been discussed theoretically in reference 23.)

The driving force for all transcrystallisation processes is the difference between the average solute concentration c_w in the matrix and the particle

* This is usually called 'Ostwald ripening' or in German 'Umlösung' which literally means 'transsolution'.

solubility c_a (see Fig. 1). So it may be useful to discuss the particle behaviour
not as a function of the particle size, i.e. with respect to an r scale, but as a
function of the particle solubility, i.e. with respect to a c_a scale, where c_a is
given by the Gibbs–Thomson equation (eqn (10)):

$$c_a = c_m \exp\left(\frac{2\gamma}{c_p RT_A \cdot r} + \frac{\Sigma E_i}{c_p RT_A}\right) \tag{15}$$

Frequently a metastable coherent α phase precipitates in the form of fine
particles at the beginning of a particular heat treatment whereas the stable

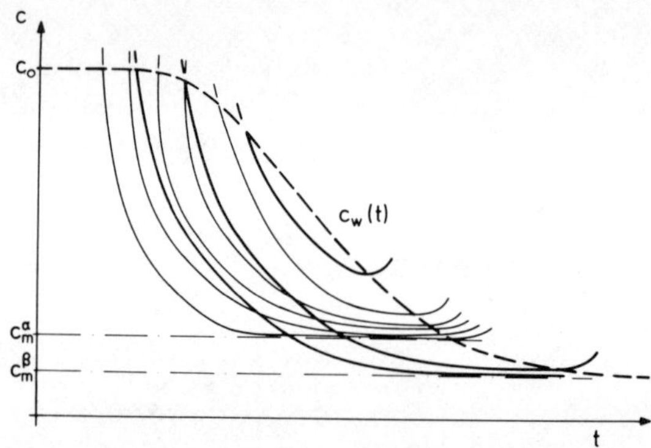

Fig. 5.   Time dependence of the particle solubilities $c_a(t)$ of phases α and β with
solubilities c_m^α and c_m^β (schematic, see text).

incoherent β phase appears only later in the form of coarse particles. Such a
precipitation process is depicted schematically in Fig. 5 in the form of a $c_a(t)$
diagram with the $c_a(t)$ curves (solid lines) of individual particles as 'growth
paths' and the $c_w(t)$ curve (broken line) describing the matrix desaturation.
As long as the particles are sufficiently small, the α particles may have a
lower solubility than the β particles (see eqn (15)). If this is so the α particles
grow faster, i.e. their 'growth paths' (fine lines) have a steeper descent, than
the already existing β particles (heavy lines). However, the equilibrium
solubility c_m^α of the α phase is the lower limit for c_a^α so that the α 'growth
paths' approach c_m^α asymptotically. This is not valid for the β 'growth paths'
so that they cross the α 'growth paths' and asymptotically approach the
smaller β solubility c_m^β. Simultaneously the concentration $c_w(t)$ in the matrix
decreases and finally becomes smaller than c_m^α so that the α particles have to
dissolve. The smaller the difference between the solubilities c_m^α and c_m^β and
the more similar the behaviour of the α and β particles (state of coherency,
diffusion coefficient) the longer both phases exist together.

It may be asked at which point does the less stable phase (α) dissolve by transcrystallisation. To answer this, assume that the matrix is already desaturated. Then the 'growth paths' reach their 'asymptotic stage' in Fig. 5 with $c_a^\alpha > c_a^\beta$, and the descent of the α curves is less steep than that of the β curves. One obtains an upper limit for the life expectancy of α if, when estimating the dissolution time of α, the coarsening rate of β is used for both phases, i.e. if the Ostwald ripening theory (Section 3.1) is used for the combined α and β distribution with the parameters for β. By means of eqn (15) the solubility of the largest α particle, to which a particular particle size $\tilde{r}_\beta$ of the β phase corresponds, can be determined. The α phase is eliminated when all particles with higher solubilities i.e. all the α particles plus the β particles with $r_\beta < \tilde{r}_\beta$, have dissolved. The fraction of the total particle number to be eliminated is obtained from the theoretical size distribution.[12] The resulting decrease in particle number corresponds to an increase in average particle size (eqn (4)) which gives the estimate of the upper limit of the α dissolution time.

A simple example for such a recrystallisation process is represented by the stress orienting of a precipitate during Ostwald ripening.[24] Instead of two phases there are two kinds of particles of one phase with two different orientations with respect to an external stress. The orientation difference means a difference in the elastic particle energy (see eqn (10)), i.e. a difference in particle solubility. The latter difference was only 0·5 % in this particular case, but it was sufficient for complete precipitate orientation by selective transcrystallisation (Fig. 6). The life expectancy of a less stable kind of particle can be estimated on the basis of the theoretical analysis of stress orientation by transcrystallisation.[13] This was confirmed by corresponding experiments.[24]

3.5 Precipitates with Inhomogeneous Distributions; Influence of Grain Boundaries and Dislocations

One or several phases may be distributed inhomogeneously because of inhomogeneities of the alloy, so that the average particle distance ζ (centre-to-centre) differs locally. The inhomogeneous structure may be characterised by the distance ζ_B between the centres of gravity of 'precipitate domains' with nearly constant ζ and likewise by ζ_P with respect to 'domains' with different phases (Fig. 7). The structure inhomogeneities are correlated with a corresponding inhomogeneous distribution of the average particle solubility so that there are compensating diffusion fluxes between different 'domains'. In other words, there is a 'domain transcrystallisation' as well as the particle transcrystallisation within the 'domains'.

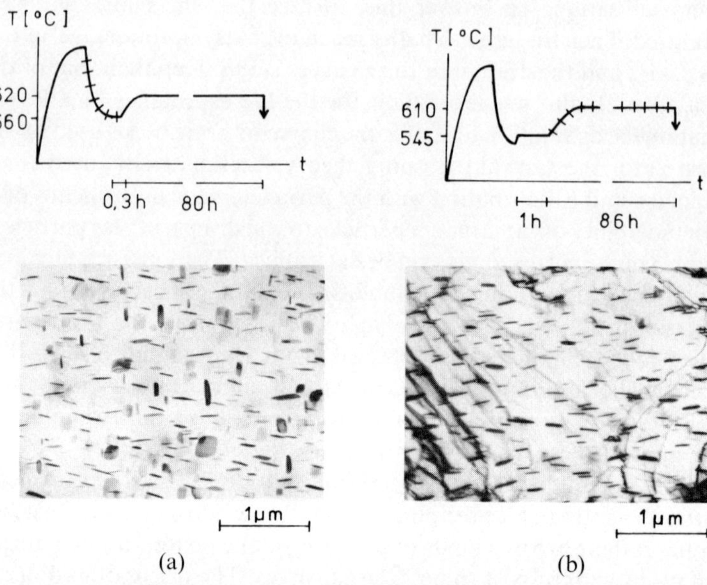

(a)   (b)

FIG. 6. Stress orientation of precipitate in a Fe–Mo–Au alloy (TEM micrographs with schematic heat treatment diagrams): (a) Little orienting after precipitation with loading only during nucleation (crossed curve); (b) complete orienting after precipitation with loading only during coarsening.[24]

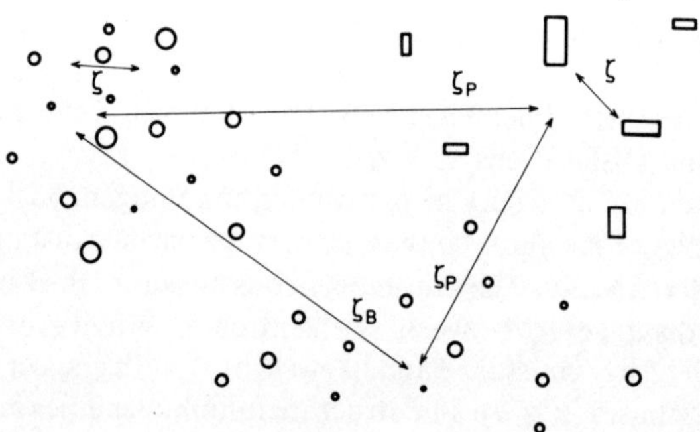

FIG. 7. Inhomogeneous distribution of two phases with (○) particles and (□) particles (schematic, see text).

To determine the structure stability, the arguments of Section 3.1 can be extended. The growth or shrinkage of 'domains' is determined by the solubilities of the particles at the 'domain boundaries'. The characteristic times $\tilde{t}_B$ or $\tilde{t}_P$ for the 'domain transcrystallisation' are appreciably longer than those for the particle transcrystallisation as a result of the longer diffusion distances ($\zeta_{P,B}^2 = 2D\tilde{t}_{P,B}$). In other words, the larger ζ_B/ζ and ζ_P/ζ are the more isolated are the domains.

Frequently the inhomogeneous particle distributions result from the influence of grain boundaries and dislocations. On one hand such lattice defects are good nucleation sites for phases with nucleation difficulties[12] so that a stable β phase is precipitated on grain boundaries and dislocations simultaneously with the precipitation of a metastable α phase between the lattice defects (see Section 3.4). On the other hand, grain boundaries and dislocations are good diffusion paths in substitutional alloys,[25] i.e. at not too high a temperature the respective diffusion coefficients D_K and D_V are larger by orders of magnitude than D_I in the defect-free lattice. However, the diffusion fluxes are determined not only by D, but also by the effective cross-section of diffusion which is $4\pi r^2$ for spherical particles in the lattice, $2\pi r\mathbf{b}$ for those on grain boundaries and $\pi\mathbf{b}^2/4$ for those on dislocations (with grain boundary thickness and dislocation diameter equal to the Burgers vector $\mathbf{b}$). Consequently the transcrystallisation on grain boundaries is determined by grain boundary diffusion only for

$$D_I \ll \frac{\mathbf{b}}{2r} D_K \tag{16}$$

The corresponding criterion for dislocation diffusion is

$$D_I \ll \frac{\mathbf{b}^2}{16r^2} D_V \tag{17}$$

If

$$D_I \gg \frac{\mathbf{b}}{2r} D_K, \frac{\mathbf{b}^2}{16r^2} D_V \tag{18}$$

the particles on defects behave in the same way as those in the defect-free lattice, and the structure stability is determined according to eqn (9) in Section 3.1. In the case of eqns (16) and (17) the changed boundary conditions of diffusion not only change τ in eqn (7), but also n.

Theoretical models for particle coarsening on grain boundaries[26-29] and on dislocations[26,29-31] can be found in the literature. $n = 4$ (eqn (7)) was consistently indicated for the coarsening on grain boundaries, and τ_K differed from τ_I by D_K—apart from geometric factors. However, it must be

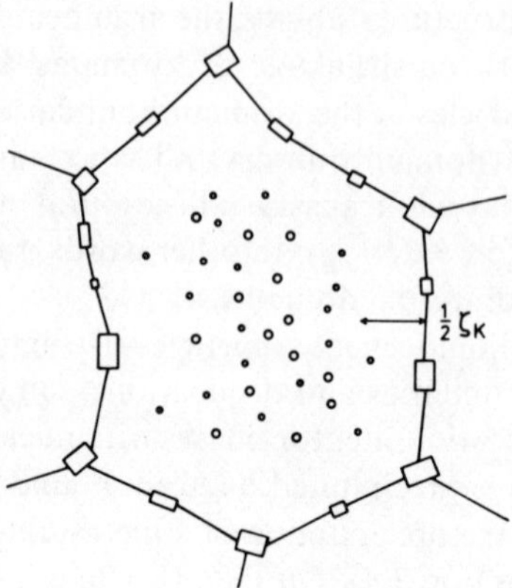

FIG. 8. Precipitation of two phases in the grain ($\bigcirc$) and on grain boundaries ($\square$) (schematic, see text).

noted that τ_K approaches zero for decreasing amounts of the precipitate phase.[28] For the coarsening on dislocations, $n = 5$ is frequently discussed,[30] whereas other detailed analyses give $n = 4$.[29,31] Thus particles on grain boundaries and dislocations coarsen faster, because of the larger diffusion coefficients, than particles in the defect-free grains, especially for small initial particle sizes r_0 and not too long coarsening intervals (small m), according to eqn (7).

To illustrate the consequences for the structure stability, it is assumed, as an example, that the precipitation of a stable phase on the grain boundaries (index K) competes with that of a metastable phase within the grains (index I) as illustrated in Fig. 8. On the grain boundary the number of nuclei is small and coarsening is fast so that there are coarse particles with lower solubilities than are found within the grain. This leads to precipitate-free zones along the grain boundaries, the extension of which can be expressed by

$$\frac{1}{2}\zeta_K \simeq \frac{\bar{r}_K}{f_K^{1/3}} \tag{19}$$

($\bar{r}_K$ = average effective radius of grain boundary particles). With increasing $\bar{r}_K$ the precipitate-free zone grows into the grain by the dissolution of the particles at its boundary until all the particles within the grain have disappeared.

3.6 A Large Amount of Precipitate

When eqns (7) and (9)—or similar ones—are used for stability discussions, the particle distances (ζ) should be large compared with the particle sizes (r) which is the case for small precipitated fractions ($f < 1\%$). For larger amounts of precipitate, e.g. duplex structures,[7] $\zeta - 2r$ and r are in the same range of order, and in superalloys with, for example, $f = 50\%$, the matrix layer between the particles is thin compared with the particle diameter.[32] The consequences for the Ostwald ripening kinetics have been discussed and a new model has been presented in reference 33. Apart from this the applicability of the Ostwald ripening theory may be questioned from a fundamental point of view[12] since in the considered case the common matrix between the particles cannot be characterised by a single average concentration c_w any longer. Instead, the local c_w is determined by the behaviour of the adjacent particles which are correlated in return with further neighbouring particles. So there is not only a local distribution of particle sizes, but also a local distribution of matrix compositions or critical particle sizes which leads to a multitude of different local coarsening processes correlated with one another.

The driving forces are not directly dependent on the precipitate fraction f nor on the coarsening topology. The diffusion paths are determined in any case by the particle distances which are proportional to the particle radii (during Ostwald ripening). Therefore the functional form of the theoretical description of coarsening must be independent of the volume fraction precipitated, and the known Ostwald ripening theory for small f is at least a valid dimensional analysis for larger f. This analysis describes the average coarsening apart from unknown numerical factors. Thus eqn (7) or eqn (8) (with the theoretical n, but initially unknown τ^*) can still be used for estimating the phase structure stability. For the particle size distribution, the logarithmic normal distribution which has been observed repeatedly (for example, reference 24) is expected, as was discussed in another context.[34]

Lamellar structures, composites and other special materials should be mentioned in this section. In such structures instabilities of phase boundaries arise leading to local shape changes of 'particles', and there is the possibility of discontinuous coarsening (or transcrystallisation). This is not discussed here, but the reader is referred to the literature.[1]

3.7 The Effect of Simultaneous Deformation

The potential retroaction of deformation processes on the simultaneous

---

* Equation (9) still gives the correct functional dependence of τ on the various parameters.

                    G. SAUTHOFF

nucleation and transcrystallisation has been investigated only sporadi-
cally,[35] and even the qualitative understanding of the partially contrasting
results is still incomplete.[36] In the following, simple criteria are given which
can be used to decide whether a retroaction of deformation on the kinetics
of the simultaneous precipitation processes, i.e. on the phase structure
stability, is expected. To simplify the discussion it is assumed at first that the
deformation results only from moving dislocations (dislocation creep).
Later there will be some comments on the opposite case of diffusion creep
without moving dislocations.

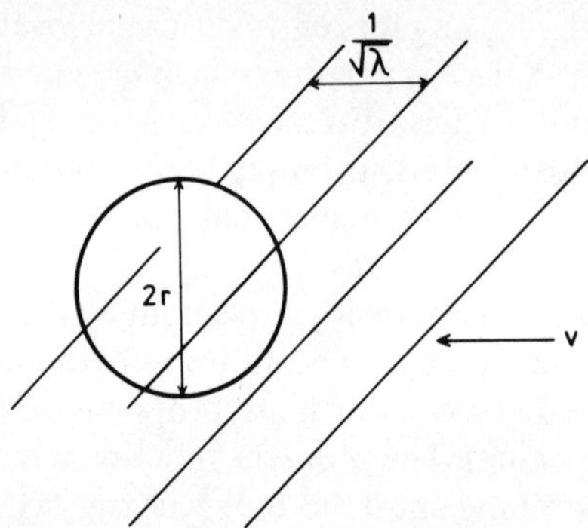

FIG. 9.   Dislocations intersecting a sphere (schematic, see text).

The dislocation creep provides additional nucleation sites on dislocations (nodes etc.) and therefore the deformation may induce an
additional nucleation in a supersaturated alloy.[37] For this to occur, the
deformation rate must be large enough in comparison with the nucleation
rate $\dot{N}$ (number of nuclei per time and volume unit) in the undeformed
alloy. The number $\dot{n}_V$ of dislocations that intersect a sphere of diameter $2r$
per unit time is in simplest approximation (see Fig. 9)

$$\dot{n}_V = \lambda . v . 2r \tag{20}$$

(λ = dislocation density and v = dislocation velocity). On the other hand,
the strain rate may be expressed by[38]

$$\dot{\varepsilon} = M^{-1} . \lambda . v . \mathbf{b} \tag{21}$$

(M = Taylor factor and $\mathbf{b}$ = length of Burgers vector). Then the rate of
dislocations that intersect a lattice site of 'diameter' $\mathbf{b}$ may be estimated by

$$\dot{n}_V = M\dot{\varepsilon} \tag{22}$$

and may be compared with the local nucleation rate $\dot{N}\mathbf{b}^3$. As long as

$$\dot{N}\mathbf{b}^3 \gg M\dot{\varepsilon}, \tag{23}$$

the retroaction of deformation on nucleation is negligible.

The later stages of growth and transcrystallisation may be influenced by concurrent deformation in various ways, with corresponding consequences for the phase structure stability. From time to time the dissolution of precipitated particles because of segregation at dislocations—for example, in Fe–C—has been discussed.[39] Such an effect can only be relevant if the segregated amount of solute is sufficiently large compared with the precipitate. If a dislocation binds φ atoms per lattice plane, the amount of segregated atoms at a dislocation (mole fraction) is

$$x_V = \varphi \, . \, \lambda \, . \, \mathbf{b}^2 \tag{24}$$

($\mathbf{b}^3 = $ lattice site volume). As long as this amount is small compared with that precipitated:

$$\varphi \, . \, \lambda \, . \, \mathbf{b}^2 \ll x_0 - x_m \tag{25}$$

(x_0 and x_m are the mole fractions in the supersaturated and saturated alloy, respectively), the influence of segregation at dislocations on the precipitate stability is negligible. For creep deformation ($\lambda \approx 10^9 \, \text{cm}^{-2}$) eqn (25) is always fulfilled in practical cases even if the comparatively large value $\varphi = 20^{39}$ is chosen.

The external stress which effects the deformation can directly influence the solubility according to eqn (10) (Section 3.2). Such solubility changes—even if they are only small, e.g. $0 \cdot 5\%$—may lead to a drastic change in the precipitate distribution as has been discussed in Section 3.4 with respect to the stress orientation of precipitates. Likewise the shape stability of precipitates may be affected by external elastic forces,[40] i.e. cubes may grow into plates or rods. The shape changes, however, do not occur by transcrystallisation, but by local diffusion processes at and in the particles. The diffusion length is characterised by the particle diameter with a corresponding short diffusion time.

Coherent and semi-coherent particles can be intersected by moving dislocations and this may lead to the shearing off of parts of the particles with much reduced sizes. However, the particle interface is increased when the particle is sheared (Fig. 10) so that smoothening diffusion processes are initiated (with the interface energy as the driving force). For the 'damping'

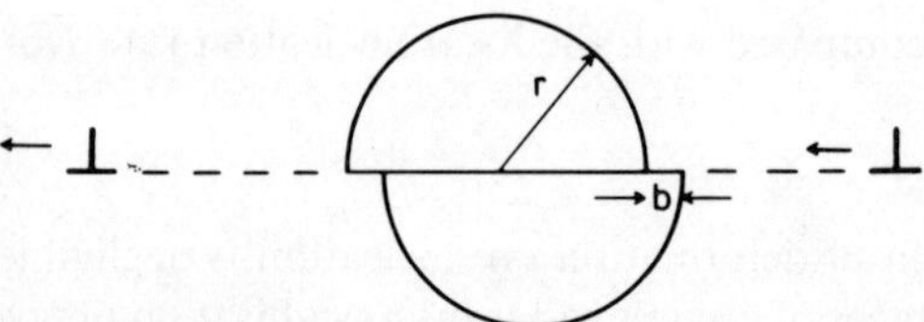

F$_{\text{IG}}$. 10.   Dislocation shearing a particle (schematic, see text).

rate $\dot{s}$ of a sinusoidal perturbation of amplitude s, and wave length S, on a plane surface it has been found that[41]

$$\frac{\dot{s}}{s} = -\frac{9}{8}\frac{1}{\tau}\left(\frac{2\pi}{S}\right)^3 \tag{26}$$

where $1/\tau$ is given by eqn (9). (For spheres a similar equation may be derived.) Therefore the smoothening rate strongly decreases with increasing wave length. In view of the shearing of particles, $S = 10\mathbf{b}$ is chosen for obtaining an estimate (in the sense of an upper limit) for the smoothening rate. Since for coarsening

$$\frac{1}{\tau} = \frac{\mathrm{d}}{\mathrm{d}t}(\bar{r})^3 \tag{27}$$

on average (see eqn (7) or reference 12), one obtains

$$\dot{s} = -\left(\frac{\bar{r}}{\mathbf{b}}\right)^2 \dot{\bar{r}} \tag{28}$$

with $s = \mathbf{b}$, $\bar{r} = $ average particle radius and $\dot{\bar{r}} = $ average coarsening rate. As long as this 'smoothening' rate is large compared with the 'shear' rate $\mathbf{b} \cdot \dot{n}_{\mathrm{V}}$ (see eqn (20)), i.e.

$$\frac{1}{2}\left(\frac{\bar{r}}{\mathbf{b}}\right)^2 \frac{\dot{\bar{r}}}{r} \gg M\dot{\varepsilon} \tag{29}$$

the effect of the particle shearing on the precipitate stability is negligible. In the local average this condition is always fulfilled under practical creep conditions ($\dot{\varepsilon} < 10^{-8}\,\mathrm{s}^{-1}$).

During deformation, the coherency of coherent and semi-coherent particles may be reduced by dislocations that arrive at the particles:[42] such dislocations may be integrated into the interface and this results in an increase in the interface energy and a decrease in the strain energy with corresponding consequences for the particle solubility according to eqn (10). However, the coherency already decreases in the undeformed material during particle growth (Section 3.2). On the basis of Brooks' criterion the rate of the spontaneous coherency decrease may be estimated by

$$\dot{n}_F = \delta \cdot \frac{2\dot{r}}{b} \tag{30}$$

which is the insertion rate of misfit dislocations with δ from eqn (11). As long as this rate is large compared with the rate of arrival of the 'deformation dislocations' (eqns (20) and (21)):

$$\frac{b}{2r}\frac{\dot{r}}{r} \gg M\dot{\varepsilon} \tag{31}$$

the effect of deformation on coherency is negligible.

As well as the driving forces which have been discussed up till now, the kinetic factors, especially the effective diffusion coefficient, may be affected by deformation. Dislocations are good diffusion paths (for substitutional atoms) and affect the particle behaviour (Section 3.5) if they are in contact with the particles for a sufficiently long time. For estimating the 'contact time' fraction, ϑ, it is assumed that a dislocation in the neighbourhood (diameter ζ) of a particle is captured and pinned for a time t_V after which it jumps to the next particle (distance ζ, see eqn (3)). The average dislocation velocity is then

$$v = \frac{\zeta}{t_V} \tag{32}$$

and with the arrival rate $\dot{n}_V$ (eqn (20)) one finds

$$\vartheta = \dot{n}_V \cdot t_V = \lambda \cdot 4r^2 \cdot f^{-2/3} \tag{33}$$

As long as this fraction is small:

$$\lambda \cdot 4r^2 f^{-2/3} \ll 1 \tag{34}$$

the direct influence of dislocations on the diffusion kinetics of trans-crystallisation is negligible.

There remains the possibility that the climbing dislocations produce a significant number of excess vacancies and thereby increase the diffusion coefficient. Since the evidence on the magnitude of this effect and on the relevant microscopic models is conflicting,[43] the problem is here considered from a macroscopic point of view. If the total deformation results only from the climb of dislocation, the plastic strain corresponds to a macroscopic diffusion flux. Therefore an equivalent diffusion coefficient $D_{\ddot{A}}$ may be defined by analogy with the description of diffusion creep:[44]

$$D_{\ddot{A}} = kT\frac{l^2}{x}\frac{\dot{\varepsilon}}{\sigma \mathbf{b}^3} \tag{35}$$

($l/\sqrt{x}$ = effective diffusion length, σ = external stress and k = Boltzmann's constant). This expression characterises the atom flux for a macroscopic strain rate $\dot{\varepsilon}$. As long as

$$D_{\ddot{A}} \approx D \tag{36}$$

with (Section 3.5)

$$D = D_{\mathrm{I}}\left(1 + \pi\frac{\mathbf{b}\,D_{\mathrm{K}}}{d\,D_{\mathrm{I}}} + \lambda\mathbf{b}^2\frac{D_{\mathrm{V}}}{D_{\mathrm{I}}}\right) \tag{37}$$

(d = grain diameter), excess vacancies are negligible. (With $l^2/x = d^2/74^{44}$ according to diffusion creep, eqn (36) represents a lower limit and states that one is within or near the diffusion creep domain in Ashby's deformation map.[3]) On the basis of a simple microscopic model[45] the increase in vacancy concentration can be estimated for a not too large value of σ by

$$\frac{\Delta C_{\mathrm{L}}}{C_{\mathrm{L}}} = \frac{1}{2}\left(\frac{\sigma\mathbf{b}^3}{kT}\right)^2 \tag{38}$$

(arithmetic mean of the vacancy concentrations at a vacancy emitting dislocation and at a vacancy absorbing dislocation). Then with

$$\frac{1}{2}\left(\frac{\sigma\mathbf{b}^3}{kT}\right)^2 \ll 1 \tag{39}$$

eqn (36) is fulfilled. A corresponding analysis of deformation experiments gives the empirical criterion[45]

$$\dot{\varepsilon}/D \leq 10^9 [\mathrm{cm}^{-2}] \tag{40}$$

These criteria are fulfilled for creep in practical cases.

In conclusion it may be stated that under creep conditions ($\dot{\varepsilon} \leq 10^{-8}\,\mathrm{s}^{-1}$) a retroaction of the dislocation creep on the precipitation processes and thus on the precipitate stability is expected only in particular cases when the nucleation or the growth is strongly inhibited or when there are several phases of (nearly) equal solubility.

In view of practical creep cases with low creep rates the possibility has to be considered that there is diffusion creep without the movement of dislocations[38] as is illustrated by Ashby's deformation maps.[3] The potential effect of diffusion creep on the structure evolution is now discussed.

In diffusion creep, e.g. in tension, atoms are detached from the grain boundaries parallel to the external stress under the influence of the local stress distribution and diffuse to the other grain boundaries (along the grain boundaries or through the grains) where they are attached again and give rise to a specimen elongation. In structures with precipitates, particles at the 'atom emitting' grain boundaries likewise dissolve. The solute atoms from the particles also diffuse following the stress gradients, but before arriving at the absorbing boundaries they precipitate at the growing particles. Therefore only solvent atoms arrive at the absorbing boundaries and form precipitate-free zones. The growth rate of such zones is equal to the grain elongation rate (order of magnitude) which for small strains up to about 10% is equal to the specimen elongation rate.[46] (For larger strains grain boundary sliding and grain switching occur so that the specimen elongates more than the grains.[46])

The diffusion fluxes that are induced by the stresses give rise to coupling stresses (see Section 3.3), and all the stresses affect the solubilities of the precipitated particles (see Section 3.2). For diffusion creep, however, these effects are negligibly small compared with the drastic structure changes at the grain boundaries.

3.8 Summary and Conclusions

Engineering materials are not in thermodynamical equilibrium; therefore they age under high temperature service conditions. The phase structure and the grain structure coarsen which usually leads to a deterioration of the mechanical properties. The arguments in Sections 3.1 to 3.7 indicate the possibilities for optimising the structure in view of sufficiently slow phase structure evolution kinetics under service conditions (i.e. 'phase structure stability') and for estimating these evolution kinetics. A 'stable' phase structure in return gives rise to a 'stable' grain structure (Section 2). Usually there is a structure with a fine dispersion of several phases which can be produced by fast nucleation processes. Thereafter the structure coarsens by transcrystallisation, the rate of which can be kept low by choosing the appropriate service conditions and alloy composition.

However, in the preceding sections it has been shown that on one hand the various physical variables which determine the evolution kinetics cannot be chosen independently of one another; on the other hand, the structure changes occur in mutual interdependence. So the measures for optimising the structure can only be aimed at the particular case in question, and there are inherent restrictions for the variation of parameters. For example, there may be the suggestion that an additional fine dispersion of particles is precipitated between coarsening particles in order to compensate for the deterioration by particle coarsening. The necessary small precipitation rate implies a strongly inhibited nucleation which usually leads to a coarse precipitate (see Section 3.1). The precipitated particles only remain fine if their growth is still much slower than their nucleation. Thus, the additional phase should be more stable than the first one by orders of magnitude.

Generally it is fast nucleation and slow growth and coarsening which favours a fine stable phase structure, i.e. the phases should have low (effective) solubilities, small interface energies and small (effective) diffusion coefficients. The kinetics of the phase changes are slower even if the diffusion in the crystal is not, but a slower interface reaction determines the rate of the phase changes. There is some evidence that the precipitation processes may be controlled by an interface reaction, but the particular mechanism is not always clear as, for example in the case of the dissolution of cementite in ferrite.[47] It seems feasible that the formation of 'growth nuclei' on the interface ('pill boxes') determines the particle growth.[48,49] With regard to coherent interfaces, a ledge mechanism has been proposed,[50] but the experimental evidence is still being evaluated.[51]

Finally the potential influence of the segregation of impurities at interfaces on the structure evolution is discussed in order to illustrate the above mechanism. Impurity atoms may segregate at the particle interfaces thus reducing the interface energy.[52] (In this way an equilibrium is established between matrix, particle and interface.) A reduced interface energy means a reduced driving force for coarsening so that segregation at phase boundaries increases the structure stability. In addition the segregation influences the kinetics of coarsening. If a particle grows faster than the impurity atoms arrive at the particle and segregate, the particle solubility increases and the growth slows down. Thus it has to be considered that the fluxes of precipitate atoms are coupled with the impurity flux. If the segregating atoms, as minority component, determine the coarsening rate, the growth equation for the particles can be derived on the basis of a mass balance as in reference 12. In the present case the particles no longer grow according to a parabolic law ($r^2 \sim t$), but according to a linear law, and the coarsening law is quadratic ($\bar{r}^2 \sim t$) instead of cubic ($\bar{r}^3 \sim t$). In other words, the precipitate behaves as one, the kinetics of which are determined by an interface reaction of the first order.[19] In a few cases the influence of segregating impurities on precipitation processes was investigated experimentally, e.g. in Fe alloys,[53] and indeed an appreciable slowing down of the precipitation by segregation was found.

REFERENCES

1. Martin, J. W. and Doherty, R. D. (1976). *Stability of Microstructure in Metallic Systems*, Cambridge University Press.
2. *Grundlagen der Wärmebehandlung von Stahl*, (Kontaktstudium II) (Ed. W. Pitsch), Verlag Stahleisen, Düsseldorf, 1976.
3. Ashby, M. F. (1973). In *Microstructure and Design of Alloys, Proc. 3rd Int. Conf. Strength of Metals and Alloys*, London, Institute of Metals, Iron and Steel, Vol. 2, pp. 8–42.
4. Lücke, K. and Gottstein, G. Ref. 2, pp. 125–42.
5. Mohamed, F. A. and Langdon, T. G. (1976). *J. Eng. Mat. Techn.*, **98**, 125–31.
6. Blum, W. (1980). In *Festigkeits- und Bruchverhalten bei höheren Temperaturen*, (Ed. W. Dahl and W. Pitsch), Verlag Stahleisen, Düsseldorf, **1**, pp. 53–99.
7. Hornbogen, E. and Kreye, H. Ref. 2, pp. 143–54.
8. Hellman, P. and Hillert, M. (1975). *Scand. J. Metallurgy*, **4**, 211–19.
9. Corti, C. W. (1978). *Scripta Met.*, **12**, 65–7.
10. Higgins, G. T. (1973). ISI/IOM meeting: Recrystallisation in the Development of Microstructures, London; (1974). *Metal Sci.*, **8**, 143–50.
11. Gleiter, H. and Chalmers, B. (1972). *Progr. Mat. Sci.*, **16**, 1–273.
12. Sauthoff, G. Ref. 2, pp. 43–55.
13. Sauthoff, G. (1976). *Z. Metallkde.*, **67**, 25–9.
14. Sauthoff, G. (1976). *Scripta Met.*, **10**, 557–9.
15. Sauthoff, G. (1975). *Z. Metallkde.*, **66**, 106–9.
16. Brown, L. M. and Woolhouse, G. R. (1970). *Phil. Mag.*, **21**(8), 329–45.
17. Kelly, A. and Nicholson, R. B. (1963). *Progr. Mat. Sci.*, **10**, 151–383 (p. 156).
18. Li, C.-Y. and Oriani, R. A. (1968). In *Oxide Dispersion Strengthening*, (Ed. G. Ansell), Gordon & Breach, pp. 431–67.
19. Wagner, C. (1961). *Z. Elektrochemie*, **65**, 581–91.
20. Björklund, S., Donaghey, L. F. and Hillert, M. (1972). *Acta Met.*, **20**, 867–74.
21. Bhattacharyya, S. K. and Russell, K. C. (1976). *Met. Trans.*, **7A**, 453–62.
22. Hillert, M. (1976). In *Physical Chemistry in Metallurgy*, (Ed. R. M. Fisher, R. A. Oriani and E. T. Turkdogan), US Steel, pp. 445–62.
23. Slezov, V. V. (1978). *J. Phys. Chem. Sol.*, **39**, 367–74.
24. Sauthoff, G. (1977). *Z. Metallkde.*, **68**, 500–5.
25. Heumann, T. Ref. 2, pp. 19–41.
26. Slezov, V. V. (1967). *Sov. Phys. Sol. State*, **9**, 927–9.
27. Geguzin, Y. E., Kaganovsky, Y. S. and Slyozov, V. V. (1969). *J. Phys. Chem. Sol.*, **30**, 1173–80.
28. Ardell, A. J. (1972). *Acta Met.*, **20**, 601–9.
29. Jain, S. C. and Hughes, A. E. (1976). AERE Report R 8415; (1978). *J. Mat. Sci.*, **13**, 1611–31.

30. Kreye, H. (1970). *Z. Metallkde.*, **61**, 108–12.
31. Slezov, V. V. and Levin, D. M. (1970). *Sov. Phys. Sol. State*, **12**, 1383–6.
32. Schubert, F. Ref. 6, **2**, pp. 211–80.
33. Brailsford, A. D. and Wynblatt, P. (1979). *Acta Met.*, **27**, 489–97.
34. Sauthoff, G. (1973). *Acta Met.*, **21**, 273–9.
35. Ridal, K. A. and Quarrell, A. G. (1962). *JISI*, **200**, 366–73.
36. Sellars, C. M. (1974). In *Creep Strength in Steel and High-Temperature Alloys*, Metals Soc., London, pp. 20–30.
37. Sauthoff, G. and Speller, W. (1976). 4th Int. Conf. Strength of Metals and Alloys, Nancy, Vol. 2, pp. 782–6.
38. Ilschner, B. Ref. 6, **1**, pp. 1–30.
39. Gridnev, V. N., Nemoshkalenko, V. V., Meshkov, Y. Y., Gavrilyuk, V. G., Prokopenko, V. G. and Razumov, O. N. (1975). *Phys. Stat. Sol.*, **31**(a), 201–10.
40. Tien, J. K. and Copley, S. M. (1971). *Met. Trans.*, **2**, 543–53.
41. Nichols, F. A. and Mullins, W. W. (1965). *Trans. AIME*, **233**, 1840–8.
42. Guichon, G., Borrelly, R. and Pernoux, E. (1975). *Phys. Stat. Sol.*, **29**(a), 601–11.
43. Ruoff, A. L. (1967). *J. Appl. Phys.*, **38**, 3999–4003.
44. Ashby, M. F. and Verrall, R. A. (1973). *Acta Met.*, **21**, 149–63.
45. Sherby, O. D. and Burke, P. M. (1967). *Progr. Mat. Sci.*, **13**, 325–90 (p. 360).
46. Crossland, I. G. and Wood, J. C. (1975). *Phil. Mag.*, **31**, 1415–19.
47. Nolfi, F. V., Shewmon, P. G. and Foster, J. S. (1970). *Met. Trans.*, **1**, 789–800.
48. Wynblatt, P. (1976). *Acta Met.*, **24**, 1175–82.
49. Ahn, T.-M., Purushothaman, S. and Tien, J. K. (1976). *J. Phys. Chem. Sol.*, **37**, 777–84.
50. Jones, G. J. and Trivedi, R. K. (1971). *J. Appl. Phys.*, **42**, 4299–4304.
51. Chen, Y. H., Doherty, R. D., Ferrante, M. and Aaronson, H. I. (1977). *Scripta Met.*, **11**, 725–42.
52. Hondros, E. D. (1969). In *Interfaces*, (Ed. R. C. Gifkins), Butterworths, London, pp. 77–100.
53. Pope, M. and Grieveson, P. (1977). *Metal Science*, **11**, 137–40.

5

Temperature and Time Dependent Transformation: Application to Heat Treatment of High Temperature Alloys

F. Schubert

KFA Julich, Julich, FRG

1. INTRODUCTION

The aim of this paper is to give an introduction to the understanding of the influence of thermo-mechanical treatments on the microstructure and properties of high temperature alloys. To obtain a complete picture, the study of references 1 to 8 is necessary.

For application temperatures above 600 °C, three groups are usually considered:

(i)   austenitic high temperature, high strength steels;
(ii)   solution hardened Ni (Co, Fe)–Cr–base alloys;
(iii) γ'-precipitation hardened Ni (Co, Fe)–Cr–base alloys.

The second and third of these groups are often called the 'superalloys'.

General rules will be given for heat- or thermo-mechanical treatment and also for the influence of long term exposure, with respect to the mechanical properties required for technical applications. For the use of the high temperature steels and superalloys in electrical power plant, in gas turbines, in chemical plant, in high temperature nuclear reactors or in aircraft turbine engines, the required properties are as follows:

(i)   high creep resistance at application temperatures;
(ii)   high resistance to thermal and mechanical fatigue;
(iii) high structural stability;
(iv) high resistance to hot corrosion in the working environment.

Within the broad field of high temperature technology, the need for high temperature, high strength materials in aircraft engines has been the main contributor to the development and the better understanding of these alloys. In an aircraft engine, Fig. 1, there are components, such as turbine blades, which must be extremely creep resistant, and components, e.g. sheet materials for combustion chambers, which have to be extremely corrosion resistant. The optimum combination of properties cannot be achieved in a single alloy, and compromise must be achieved by alloy composition and thermo-mechanical treatment. The influence of heat treatment on properties is closely connected with microstructural features. Beginning with the chemical composition, the as-received or starting structure of the alloys estimated by the heat treatment, are described and their stability during long term exposure is discussed.

2. STRUCTURE OF HIGH TEMPERATURE STEELS AND SUPERALLOYS

2.1 Chemical Composition (Table 1(a) and (b))*

The high temperature austenitic steels have been developed from the corrosion resistant 18% Cr–8% Ni stainless steels. The higher stability of the austenitic phase at high temperatures is obtained by increasing the nickel content and reducing the chromium content to about 16%. The addition of niobium and/or titanium has an effect similar to that in corrosion resistant steels: the carbon precipitates out as monocarbides and the chromium denudation at grain boundaries is reduced.

In solid solution, molybdenum and/or tungsten increase the high temperature creep resistance. Low boron content, due to the segregation of the boron atoms at the grain boundaries and at dislocation networks, reduces the defect fraction and causes an improvement in creep resistance. In Table I(a) and (b) a few, currently used austenitic steels are listed.

By replacing some of the iron by increasing nickel additions, higher levels of chromium can be obtained in the austenitic γ-region with the benefit of higher corrosion resistance, as indicated in Fig. 2.

The solid solution hardened Ni–Cr, Ni–Cr–Co and Co–Cr–Ni alloys are closely related to the austenitic steels and the strengthening mechanisms are very similar. These mechanisms include:

(i)   hardening by carbide precipitation;
(ii)   solid solution hardening by molybdenum and tungsten;

---

* In this paper, as well as the German DIN-termed names, the well-known internationally used names will be used.

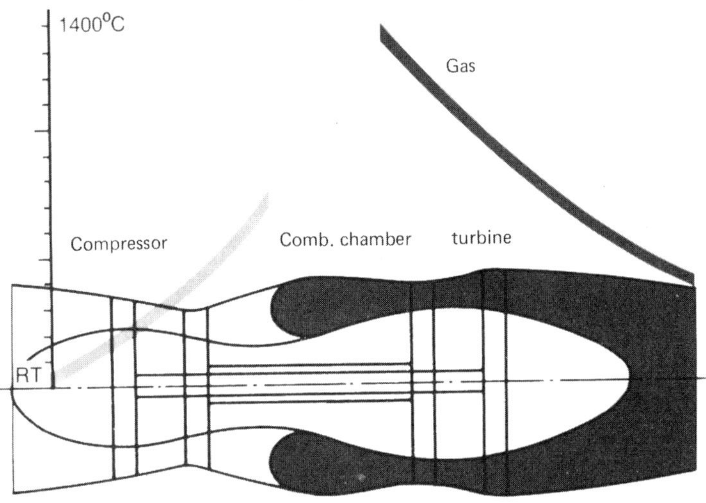

Fig. 1. Temperature profile in a gas turbine (schematic). Turbine blade: $R_{m/5000} = 210\,\text{N/mm}^2$; turbine disc: $R_{p0\cdot2} = 500\,\text{N/mm}^2$; turbine vane: residual $0\cdot2\%$, yield strength, $17\cdot5\,\text{N/mm}^2$; Combustion Chamber: hot gas corrosion resistance residual properties.

(iii)   the addition of boron for grain boundary strengthening;

(iv)   the addition of rare earth elements (such as La, Ce) to improve the corrosion resistance.

In addition the elements Si, Mn, Al and Ti are of great importance for the stability of surface oxide layers.

If the amounts of Al and Ti exceed the solubility limit, Fig. 3, hardening by precipitation of γ' [Ni_3, (Al, Ti, Nb), type $L1_2$] is possible. All such alloys have compositions within the two phase region in the phase diagrams, carefully balanced to achieve the desired properties. In the γ'-phase, the Al can be replaced by Ti and/or Nb only to a certain extent, e.g. Ni_3Ti is not stable at higher temperatures or during long term exposure above 650 to 700 °C, and the stable phases η (Ni_3Ti) or δ (Ni_3Nb), are often precipitated as needle-shaped particles which have a detrimental effect on some mechanical properties. Representative of this group are the well-known alloys A 286 and Inconel 718. The last group of Table 1, with increasing amounts of γ'-forming elements, which is equally an order of increasing creep rupture strength at high temperature, is divided into wrought and cast

TABLE 1

NOMINAL MECHANICAL COMPOSITION OF SOME HIGH TEMPERATURE, HIGH STRENGTH AUSTENITIC STEELS AND SUPERALLOYS

a) Austenitic steels and solid solution hardened superalloys

No.	Name	C	Fe	Ni	Cr	Co	Mo	W	Nb	Ta	Ti	Al	Zr	B	Mn	Si	Others	
1	X6Cr Ni 18 11	0·06	balance	11	18	—	—								+	+		~AISI 304
2	X6Cr Ni 17 13	0·06	balance	13	17	—	2·2								+	+		
3	X8Cr Ni Mo Nb 16 16	0·06	balance	16·5	16·5	—	1·8	—	0·7					+	+	+		
4	X6Cr Ni W Nb 10 16	0·06	balance	16·5	16·5	—	—	3	1					+	+	+	N 0·1	
5	X12Cr Co Ni 21 20	0·12	balance	20	21	20	3	2·5	1					+	+	+	N 0·15	
6	X8Ni Cr 32 22	0·08	balance	32·5	21	—	—	—	—	—	0·4	0·4	—	—	0·4	0·8	—	Incoloy 800
7	Hastelloy X	0·10	18·5	47	22·0	15	9·0	0·6	—	—	—	—	—	—	0·50	0·50	—	1
8	Inconel 625	0·05	2·5	balance	21·5	—	9·0	—	3·6	—	0·2	0·2	—	—	0·20	0·02	—	2
9	Inconel 617	0·07	—		22·0	12·5	9·0	—	—	—	—	1·0	—	—	—	—	—	3
10	Haynes 188	0·10	1·5	22·0	22·0	balance	—	14·0	—	—	—	—	—	—	0·60	0·40	0·08 La	4

+ : Metallurgically influenced minor contents.

b) *γ'-Hardened superalloys*

No.	Name	C	Fe	Ni	Cr	Co	Mo	W	γ'-forming elements				Zr	B	Mn	Si	Others	Amount
									Nb	Ta	Ti	Al						
1	A 286	0·05	balance	26·0	15·0	—	1·3	—	—	—	2·0	0·2	—	0·015	1·35	0·50	—	5
2	Inconel 718	0·04	18·5	balance	19·0	—	3·0	—	5·0	—	1·0	0·5	—	—	—	—	—	6
3	Nimonic 60 A	0·05	—	balance	21	—	—	—	—	—	2·5	1·5	0·05	0·005	0·10	0·20	—	7
4	Nimonic 90	0·07	—	balance	19·5	16·5	—	—	—	—	2·5	1·5	0·05	0·005	—	0·20	—	8
5	Waspaloy	0·05	—	balance	20·0	15·0	4·3	—	—	—	2·7	1·5	0·05	0·005	—	—	—	9
6	Udimet 520	0·05	—	balance	19·0	12·5	6·0	1·0	—	—	3·2	2·0	—	0·005	—	—	—	10
7	Udimet 710	0·07	—	balance	18·0	15·0	3·0	1·5	—	—	5·0	2·5	0·07	0·020	—	—	—	11
8	IN 713 LC	0·05	—	balance	12·0	—	4·5	—	2·0	—	0·5	6·0	0·10	0·010	—	—	—	12
9	IN 718 LC	0·15	—	balance	16·0	9·0	1·7	2·6	1·0	1·7	3·5	3·5	0·10	0·010	—	—	—	13
10	IN 100	0·15	—	balance	10·0	15·0	3·0	—			4·7	5·5	0·05	0·015	—	—	V 1·0	14
11	Mar M 200 DS	0·12	—	balance	9·0	10·0	—	12·0	1·0	—	2·0	5·0	0·05	0·015	—	—	Hf + 1·4	15

F. SCHUBERT

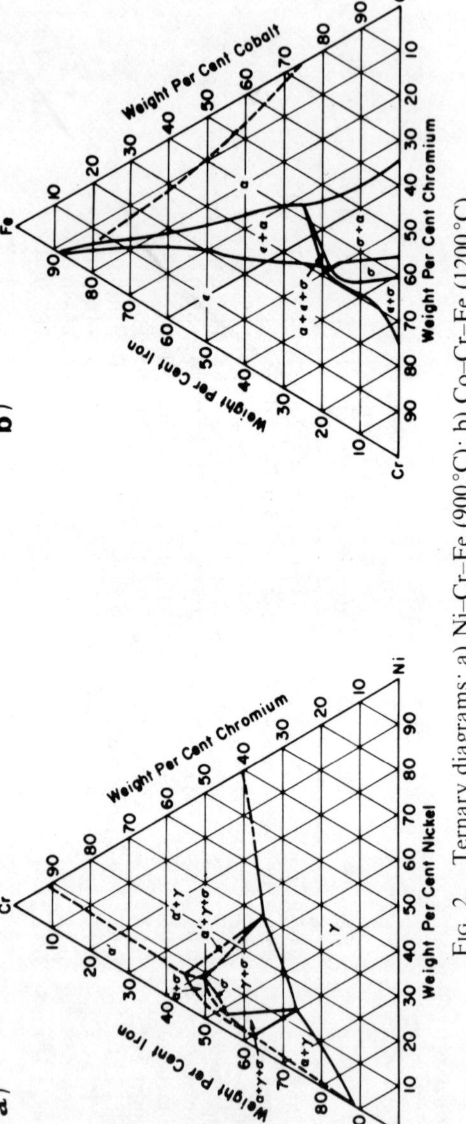

Fig. 2. Ternary diagrams: a) Ni–Cr–Fe (900 °C); b) Co–Cr–Fe (1200 °C).

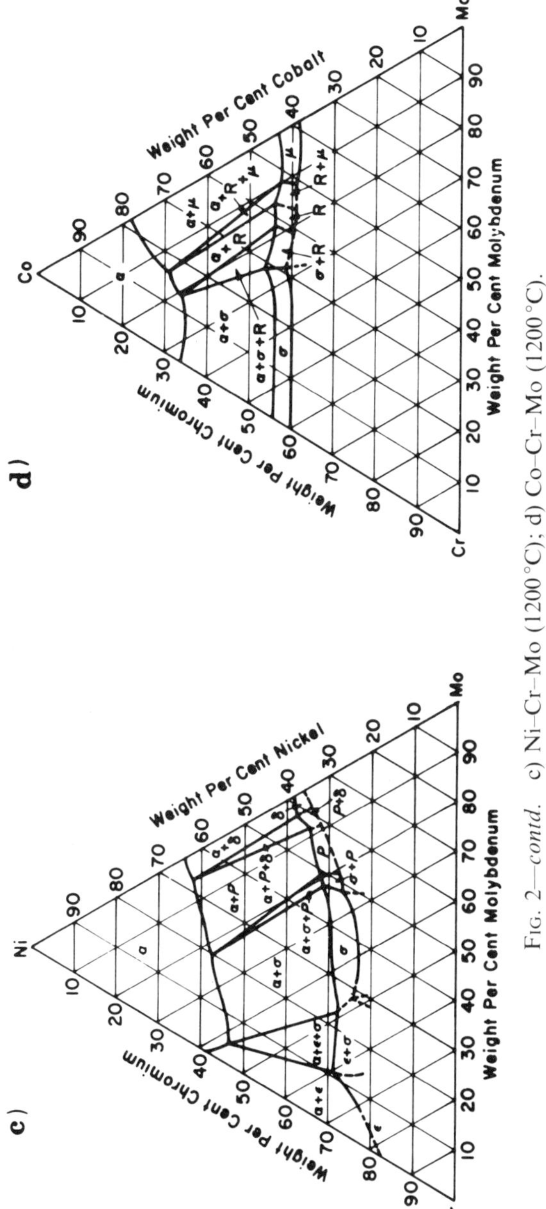

Fig. 2—contd. c) Ni–Cr–Mo (1200°C); d) Co–Cr–Mo (1200°C).

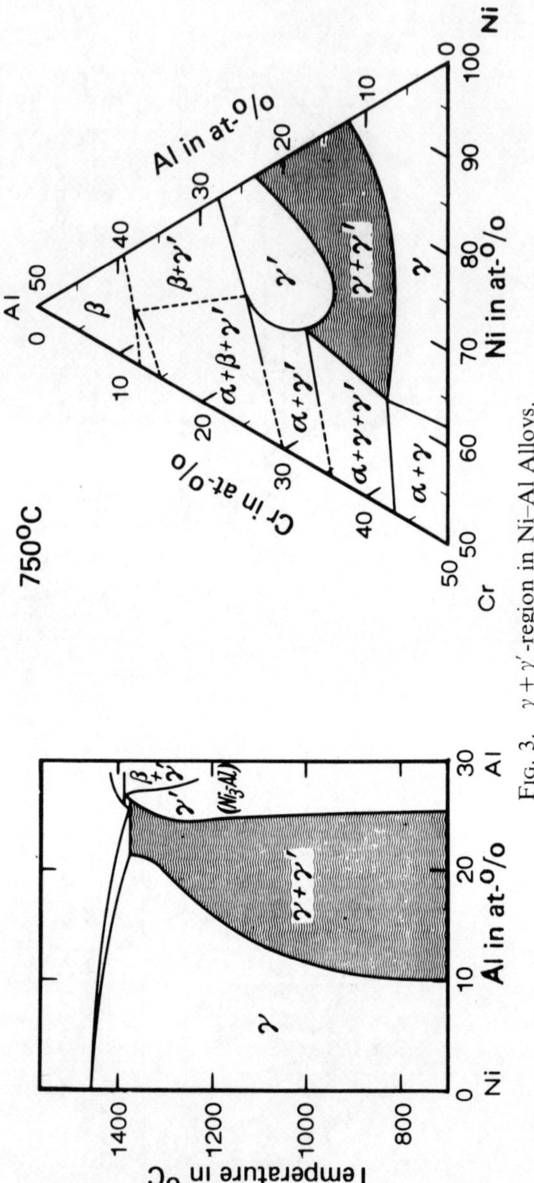

Fig. 3. $\gamma + \gamma'$-region in Ni–Al Alloys.

800°C
Ni
Al in at-%
Al
Ni in at-%
Nb in at-%
Nb
γ
γ+γ'
γ'
γ'+δ
γ+γ'+δ
γ+δ
δ
750°C
Ni
Al in at-%
Al
Ni in at-%
Ti in at-%
Ti
γ
γ+γ'
γ'
γ+η
γ+γ'+η
γ+η

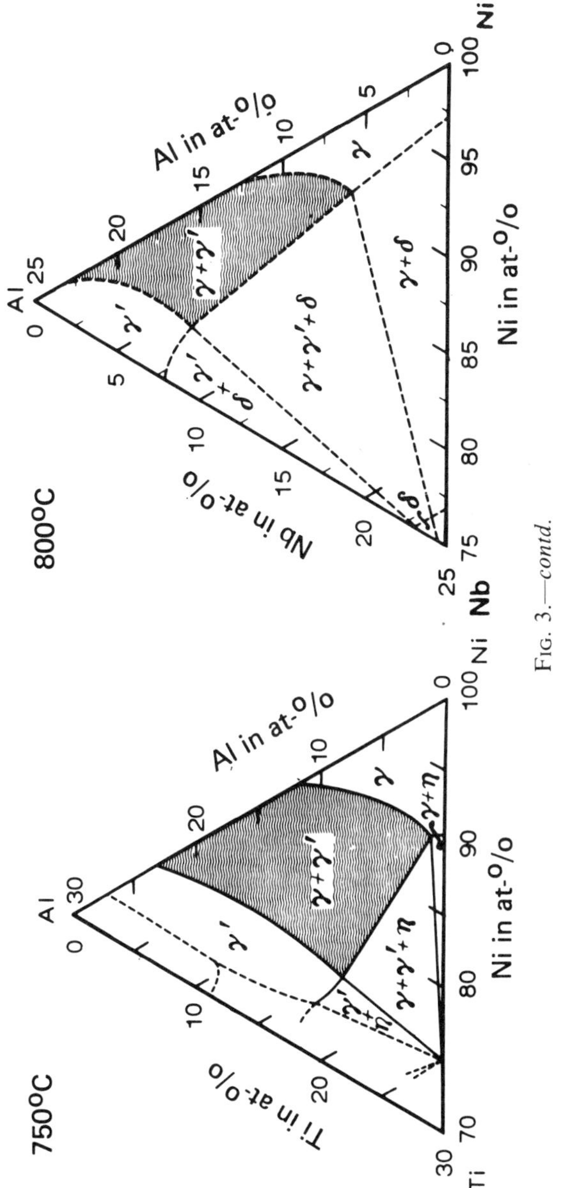

Fig. 3.—*contd.*

F. SCHUBERT

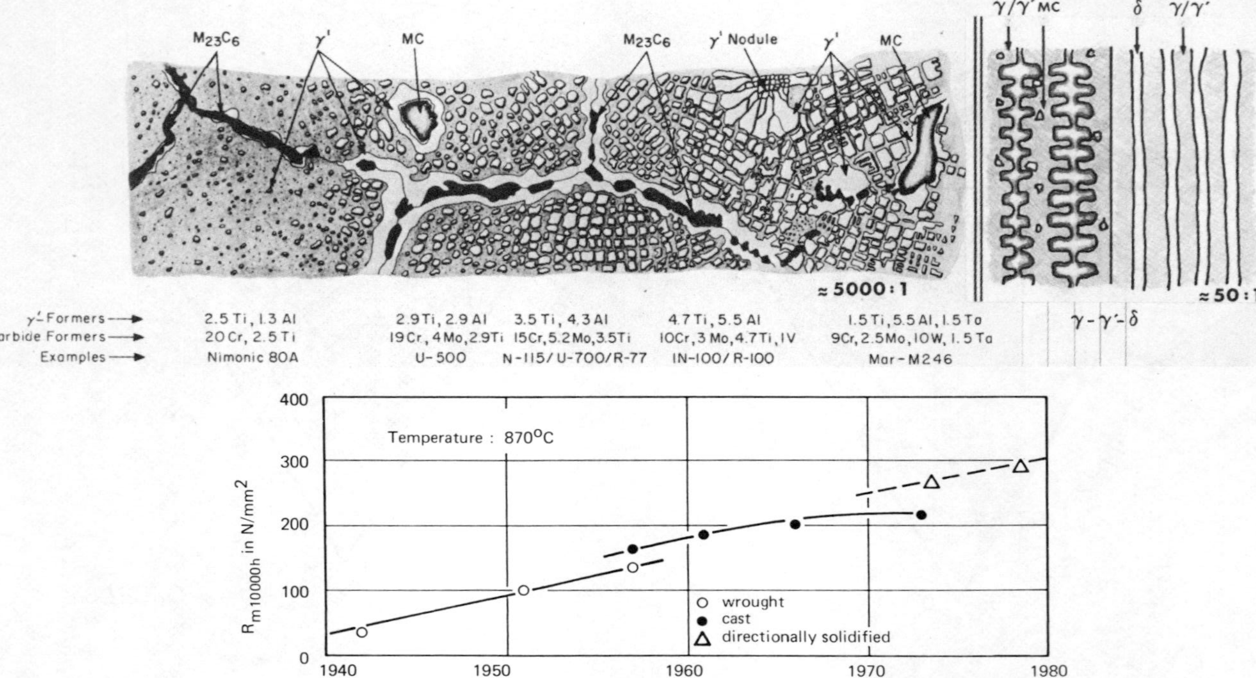

Fig. 4. Development of high temperature Ni-base alloys. Year of introduction and microstructural features.

alloys. Those alloys with more than 45% vol. of γ', are conventionally formed to components only by vacuum investment casting methods, and are not amenable to hot and cold forming.

2.2 Microstructure

The rather complex composition of the alloys is designed to yield a critically balanced 'initial' structure consisting of several phases. The phases which may occur and their characteristic microstructural features are demonstrated schematically in Fig. 4, which shows the historical development of nickel-base superalloys. The structure of austenitic steels and solid solution strengthened superalloys start on the left side of this diagram. The phase detected in common, high temperature steels and superalloys (the initial structure, i.e. in as-cast or as-heat-treated condition), are:

(i)   a solid solution hardened fcc matrix;
(ii) a dispersion of carbides (MC, $M_{23}C_6$, M_6C etc.);
(iii) geometrically close packed phases (γ').

While γ'-strengthened Fe- and Ni-base alloys are used, similarly strengthened Co-base alloys are not, due to the high temperature instability of γ' in Co-base material. In some of the alloys M_3B_2 type precipitates and some carbonitrides are present. Small amounts of undesired phases (sulphides, complex carbo-sulphides and others) may occur.

Whereas for austenitic steels and solid solution hardened superalloys, the final heat treatment is to control the grain size, which is dependent on the volume fraction of primary carbides, e.g. TiC, the task of the heat treatment of superalloys is to control the grain size and secondly to control the size and distribution of γ'-precipitation. Some cast superalloys with high γ'-volume fractions are used in as-cast condition, and the microstructures are then strongly influenced by the casting process parameters.

3.   HEAT TREATMENT

To achieve an optimal microstructure for a specific application the heat treatment should consist of:

(i)   a solution heat treatment stage; and
(ii) precipitation ageing treatment stages.

An approximate TTT diagram will be helpful in understanding the

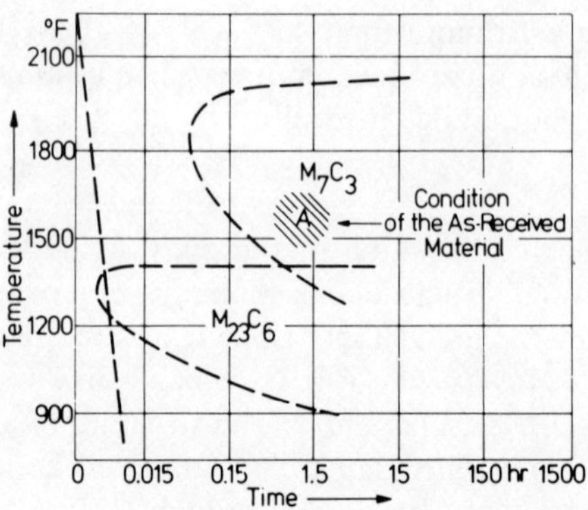

Fig. 5.   Tentative TTT diagram for IN-600.

influence of time and temperature on microstructures, although for extended ageing times, diagrams are only available for a few superalloys.

3.1   Austenitic Steels and Solid Solution Hardened Superalloys

As an example of a solid solution strengthened superalloy, which also applies for austenitic steels, a tentative TTT-diagram is given[9] in Fig. 5, for

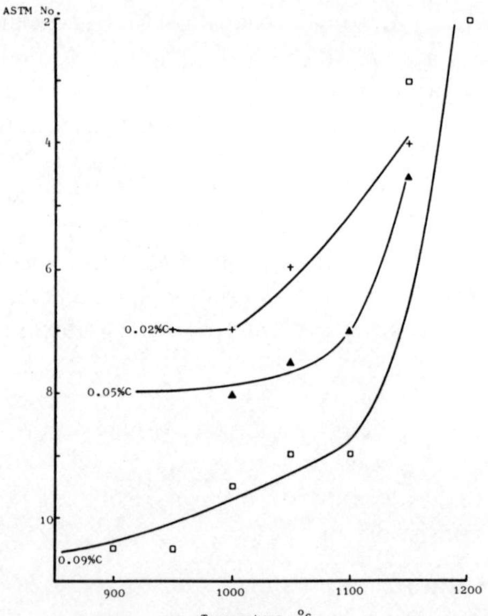

Fig. 6.   Grain size dependence on annealing temperature and C-content of Alloy 800.

a 0.7% C–16.6% Cr–7.9% Fe–bal. Ni alloy. Theoretically, a solution heat treatment means annealing at high temperature to dissolve all precipitates and holding at this temperature to produce the required grain size. In the austenitic alloys, which contain no monocarbide forming elements, the cooling rates after solution treatment must be very high to avoid precipitation of $M_{23}C_6$ carbides. The mechanical properties of niobium- or titanium-free steels are mainly influenced by the precipitation of carbides, e.g. $M_{23}C_6$, at grain boundaries, twin-boundaries and dislocation tangles.

In austenitic steels and solid solution strengthened alloys, the grain size increases with increasing heat temperature. This relationship is not linear, especially in those alloys where the grain size is also controlled by the amount of monocarbide forming elements, e.g. titanium, and by the degree of preceding cold deformation, Fig. 6.[10] In alloys (such as Alloy 800) dependent on the volume fraction of TiC, grain coarsening occurs at temperatures between 900 °C and 1250 °C. To produce a fine grain size with higher strength, the annealing temperature should be sufficiently low for all the carbon to remain tied up in the TiC.

3.2 γ'-Strengthened Superalloys

A quasi-binary schematic phase diagram of the sum of solid solution elements/sum of γ'-forming elements, Fig. 7, will make clear the problems of heat treatment.[11] In practice, the final heat treatment has two major parts:

(i) Annealing at a high enough temperature for the alloy to be in homogeneous solid solution with the aim of (a) bringing into solution those γ'-precipitates produced during hot forming, (b) homogenising the γ'-forming elements in the solid solution, and (c) controlling the grain size. Carbides of the $M_{23}C_6$-type can be made to dissolve. The temperature required increases from about 950 °C up to 1200 °C with increasing amounts of γ'-forming elements, i.e. with increasing percentage volume fraction of γ'.

(ii) Annealing at temperatures in the heterogeneous γ/γ'-phase region with the aim of controlled γ' precipitation in regard to fraction, size, morphology and distribution. These structural features are dependent on temperature, holding time at temperature, cooling rates and number of heat treatment stages.

For alloys with a γ'-fraction of about 20% vol. (e.g. Nimonic 80 A), the normal heat treatment has two stages:

(i) Solution annealing: dissolution of γ' and $M_{23}C_6$-precipitates, recrystallisation of γ solid solution and control of grain size.

                    F. SCHUBERT

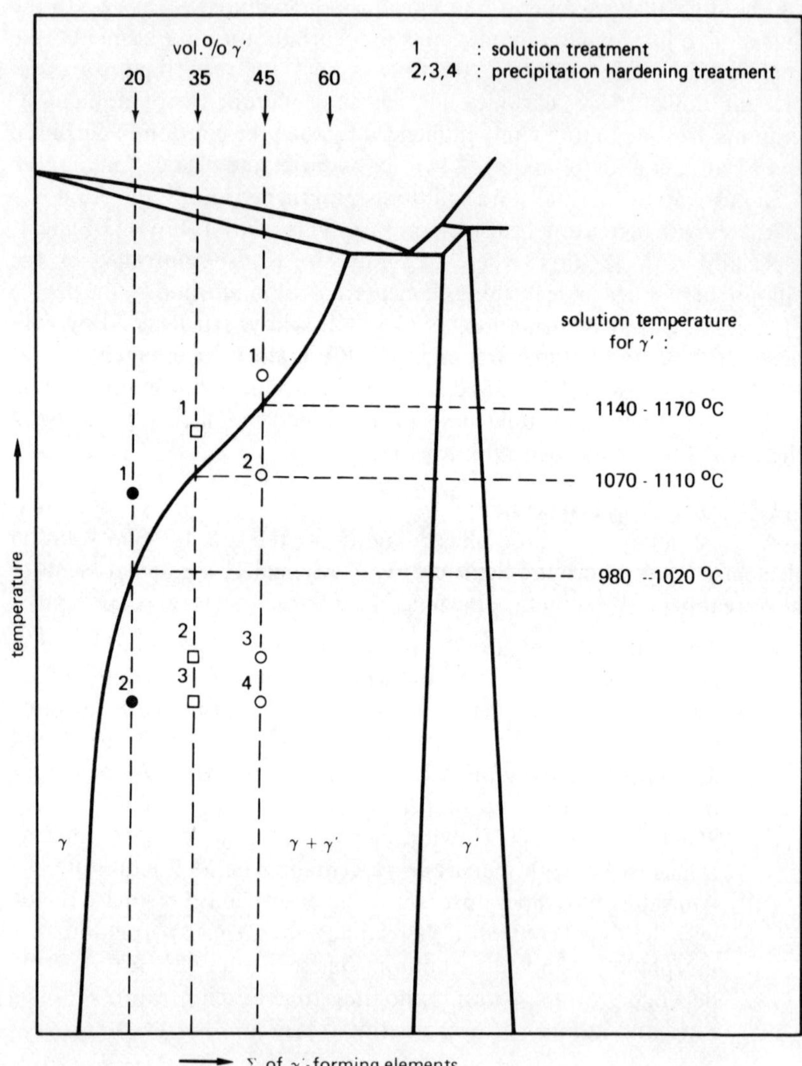

F IG. 7.   Scheme for heat treatment for γ'-hardened superalloys.

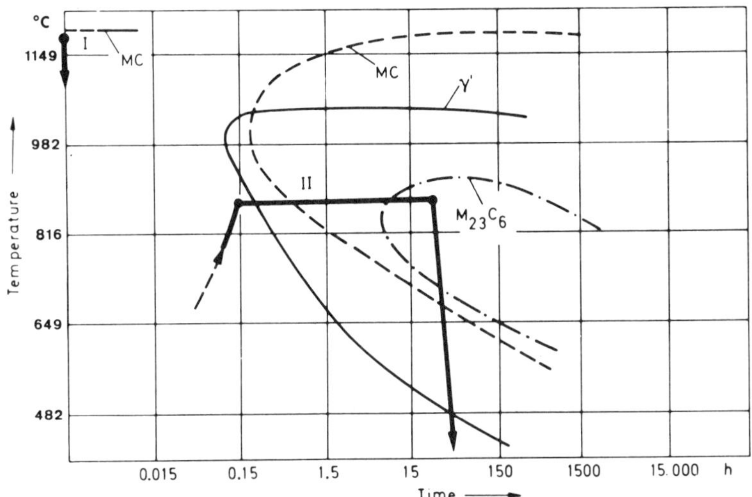

Fig. 8. TTT diagram and heat treatment cycle for Ni-base alloy IN-700 (0·1 % C–16·0 % Cr–44·0 % Ni–3·0 % Mo–29·5 % Co–2·5 % Ti–3·0 % Al).[12]

(ii)   Ageing annealing: precipitation of γ' homogeneously distributed within the grains and of $M_{23}C_6$ at grain boundaries and twin-boundaries.

An example of this kind of two step treatment for an alloy with the nominal composition 0·1 % C–16 % Cr–3 % Mo–30 % Co–2·5 % Ti–3 % Al–bal. Ni is given in the TTT diagram of Fig. 8.[12]

In alloys of this kind, the mechanical properties should be strongly dependent only on the particle size. The room temperature properties, such as hardness, correspond to the medium partial size diameter, Fig. 9, which may be achieved by different combinations of temperature and holding time.[13]

Alloys with a percentage volume fraction of about 30 %, like Waspaloy, Udimet 500 and Udimet 520, are commonly heat treated in three stages:

(i)   Solution treatment: as in the two-stage procedure.
(ii)   First ageing annealing: precipitation of γ', dependent on the combination of holding time and temperature; precipitation of $M_{23}C_6$.
(iii) Second ageing annealing: increasing the amount of γ'-precipitates.

By this kind of heat treatment, the structure of the alloy contains homogeneously distributed globular γ'-particles with a medium size

F. SCHUBERT

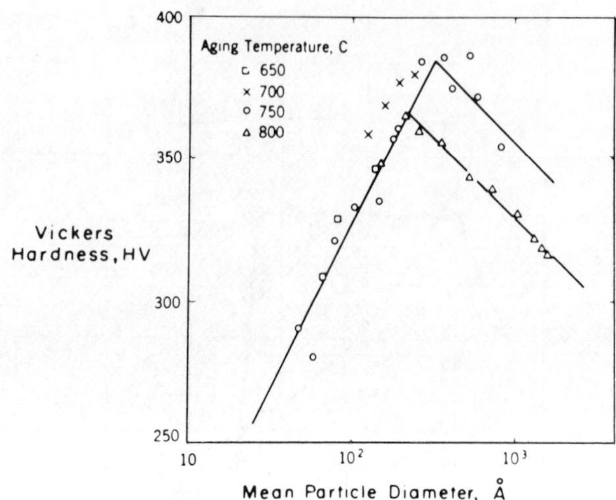

FIG. 9. Dependence of room temperature hardness of a Ni–Cr–Al alloy on ageing treatment (22 % Cr–2·8 % Ti–3·1 % Al).

diameter of about 300 Å. Using a higher temperature for the first ageing annealing, Fig. 10, (heat treatment: 1120 °C/4 h/air + 925 °C/4 h/air + 760 °C/16 h/air) the medium particle size of a Udimet 520 specimen[11] is somewhat greater than if the 845 °C, first ageing annealing treatment were used. A comparison of the mechanical properties after the two heat treatments of the same material, indicates that for room temperature and short term creep properties the finer γ' particles produced by the 845 °C treatment have some advantages. On the other hand, the 925 °C treatment which results in a larger particle size, increases the creep rupture strength for application exposures at higher temperatures.

To some extent, a γ'-strengthened superalloy should, after similar heat treatment, have reproducible microstructure and mechanical properties, but scatter is introduced due to the residual influences of the distribution of carbide precipitation and of the thermo-mechanical treatment.

If the final hot working is performed at temperatures high enough above the γ'-solution temperature, in the region of homogeneous solid solution, depending on deformation rates, a dynamic, recrystallised grain structure is obtained.[14] A solution annealing results in an increase in grain size. At the final hot forming temperature just below the γ'-solution temperature, however, dynamic recrystallisation is hindered to a certain extent by the beginning of γ'-precipitation with the result that in the microstructure, a residual amount of energic deformation is retarded. By applying a post

(a)

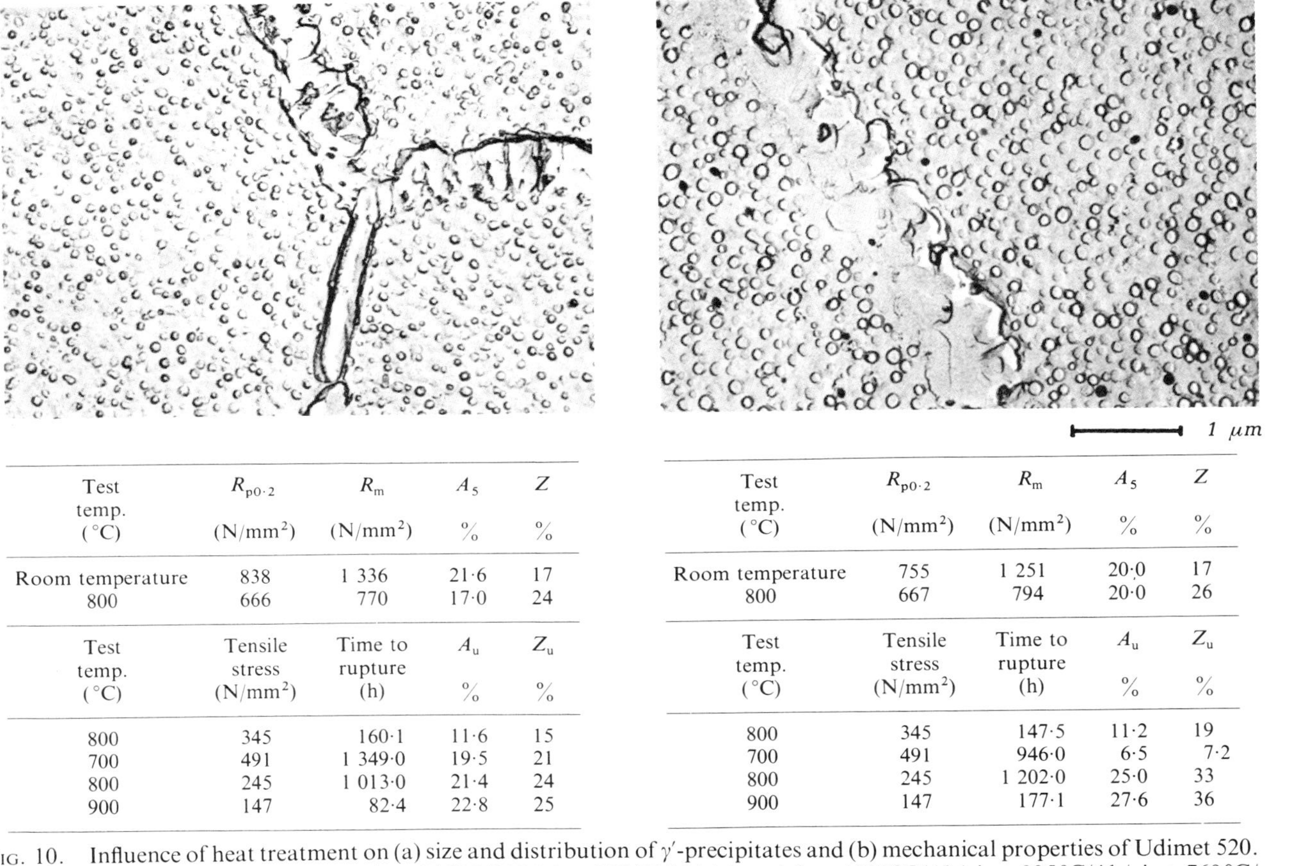

(b)

Test temp. (°C)	$R_{p0\cdot2}$ (N/mm²)	R_m (N/mm²)	A_5 %	Z %
Room temperature	838	1 336	21·6	17
800	666	770	17·0	24

Test temp. (°C)	Tensile stress (N/mm²)	Time to rupture (h)	A_u %	Z_u %
800	345	160·1	11·6	15
700	491	1 349·0	19·5	21
800	245	1 013·0	21·4	24
900	147	82·4	22·8	25

Test temp. (°C)	$R_{p0\cdot2}$ (N/mm²)	R_m (N/mm²)	A_5 %	Z %
Room temperature	755	1 251	20·0	17
800	667	794	20·0	26

Test temp. (°C)	Tensile stress (N/mm²)	Time to rupture (h)	A_u %	Z_u %
800	345	147·5	11·2	19
700	491	946·0	6·5	7·2
800	245	1 202·0	25·0	33
900	147	177·1	27·6	36

FIG. 10. Influence of heat treatment on (a) size and distribution of γ'-precipitates and (b) mechanical properties of Udimet 520. Heat treatments; left: 1100°C/4 h/air + 845°C/24 h/air + 760°C/16 h/air; right: 1120°C/4 h/air + 925°C/4 h/air + 760°C/16 h/air.

                                   F. SCHUBERT

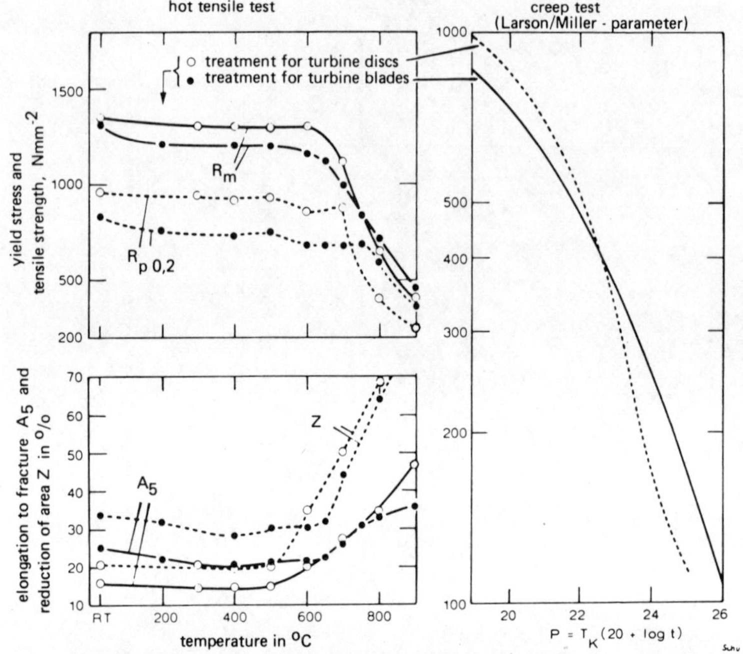

Fig. 11. Influence of different treatments on; left, the tensile properties of Waspaloy; and right, the Larson–Miller plot of Waspaloy.

solution treatment to prevent an increase in the γ-solution temperature, this kind of hot or hot/cold working can be used. For Waspaloy, both kinds of thermo-mechanical, or heat, treatment are used, the first one for turbine blade, the second for turbine disc, applications.[14]

Comparison of the thermo-mechanical and heat treatments, (Fig. 11) shows a higher, 0·2%, yield and rupture strength, up to temperatures of about 750 °C, for the thermo-mechanical treatment, which benefits from work hardening; however, a loss of creep rupture strength at higher temperature or lower stress levels is obtained with this treatment.

For alloys containing a γ'-fraction of about 40 to 45% vol., e.g. Udimet 700 or Udimet 710, a four-stage heat treatment is commonly used, Fig. 12:[11,12]

(i)   solution annealing at temperatures higher than 1100 °C;

(ii)   first ageing annealing at temperatures as high as 1080 °C;

(iii) second ageing annealing for stabilisation of the γ'-precipitations and to build precipitates of fine γ' between the former particles; carbides may also be precipitated;

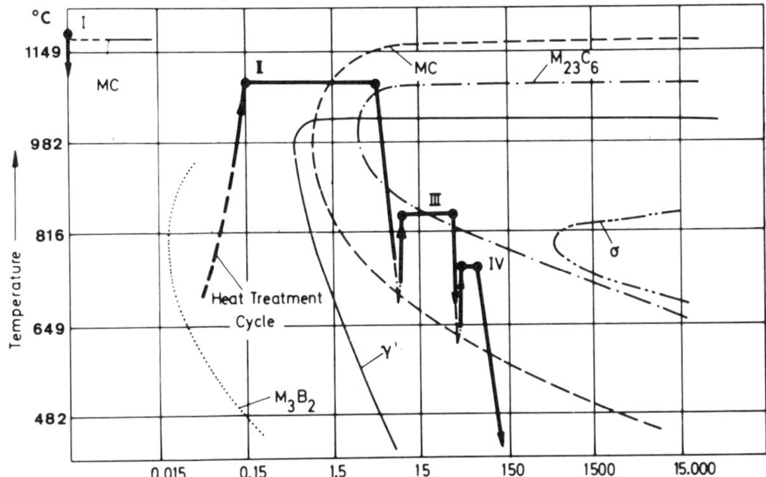

Fig. 12.   TTT diagram and heat treatment cycle for Ni-base alloy Udimet 700 (0·7%C–15·0%Cr–52·7%Ni–5·3%Mo–18·5%Co–3·5%Ti–4·3%Al–0·03%B).[12]

(iv)   third ageing annealing to increase the amount of fine γ'-precipitation.

The sequence of microstructures obtained in the different heat treatment steps are given in Fig. 13, for Udimet 710.[11]

γ'-Strengthened Ni-base superalloys, produced by vacuum investment casting, are dependent on increasing γ'-content, either two- or three-stage heat treated or used in the as-cast condition. For the microstructural features and therefore also for the mechanical properties, the following process parameters are of importance:[15]

(i)   superheating temperature of the liquid melt;
(ii)   pouring temperature;
(iii)   shell mould temperature;
(iv)   metal-mould-equilibrium (MME) temperature;
(v)   time taken to reach the MME point;
(vi)   cooling rate;
(vii)   heat flow.

All these parameters have a strong impact on microstructural features, such as grain size and orientation, dendritic arm spacing, coalescence and γ'-distribution; they also govern the mechanical properties, Fig. 14. A subsequent heat treatment cannot modify or eliminate all these influences.

F. SCHUBERT

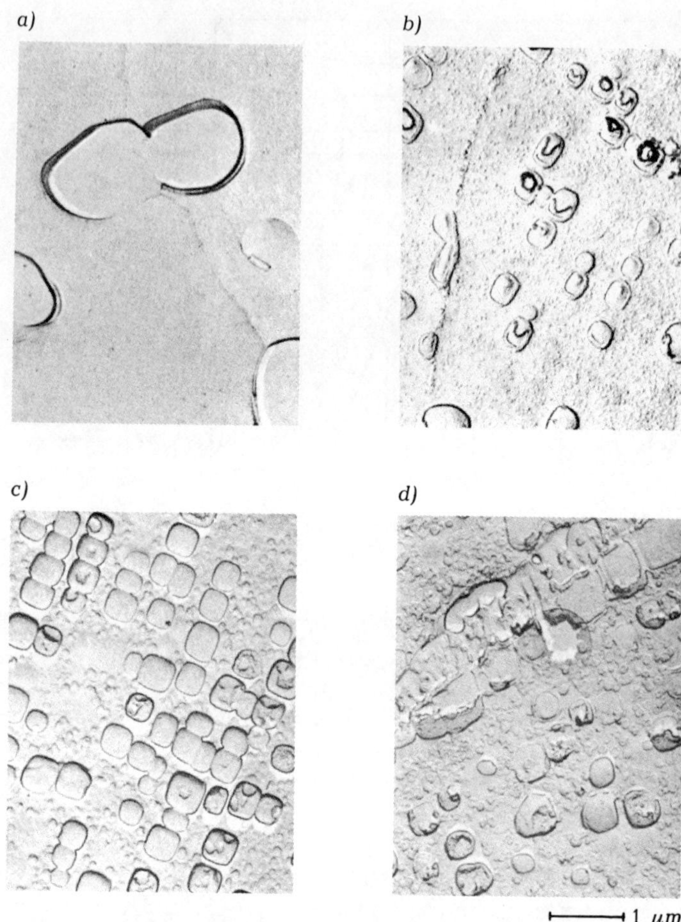

a)   b)   c)   d)

FIG. 13. Influence of heat treatment steps on the microstructure of ATS 382 alloy ($\approx$Udiment 710): (a) 1175°C/4 h/Water Quenched; (b) +1080°C/4 h/Water Quenched; (c) +845°C/24 h/Water Quenched; (d) +760°C/16 h/Water Quenched.

In addition to being influenced by the parameters of thermo-mechanical treatments the morphology of the coherent γ'-phase also depends on the composition and the lattice parameter mismatch between the a_γ matrix and $a_{\gamma'}$, Fig. 15.[16] By changing the content of molybdenum as well as the aluminium to titanium ratio, Loomis[16] demonstrates, for a series of experimental alloys, the different kinds of morphology; spheroidal, globular, blocky and cuboidal, with increasing $a_\gamma/a_{\gamma'}$ mismatch. By altering

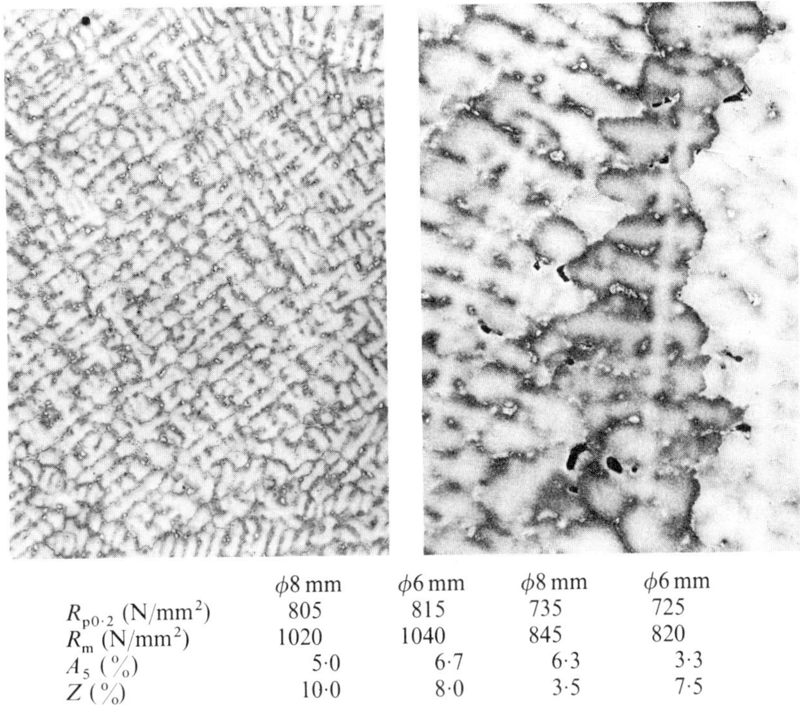

	$\phi 8\,\text{mm}$	$\phi 6\,\text{mm}$	$\phi 8\,\text{mm}$	$\phi 6\,\text{mm}$
$R_{\text{p0·2}}$ (N/mm^2)	805	815	735	725
R_{m} (N/mm^2)	1020	1040	845	820
A_5 (%)	5·0	6·7	6·3	3·3
Z (%)	10·0	8·0	3·5	7·5

FIG. 14.   Influence of casting parameters on structure and mechanical properties of IN 738. Left: rapid solidification. Right: slow solidification. Scale 1 mm: 0·029 mm. Room temperature properties (as cast).

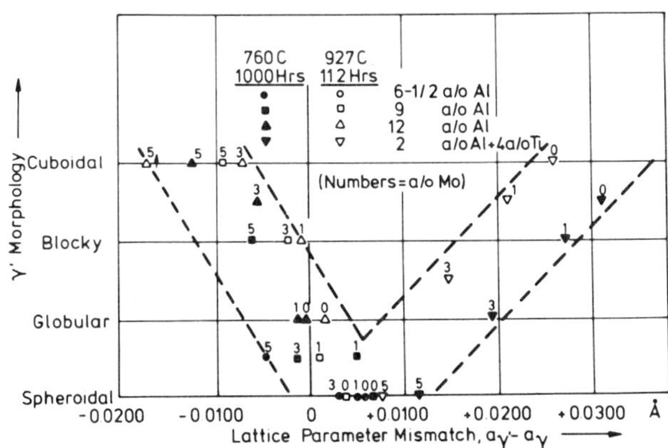

FIG. 15. γ'-Morphology and lattice parameter mismatch of a Ni-superalloy.

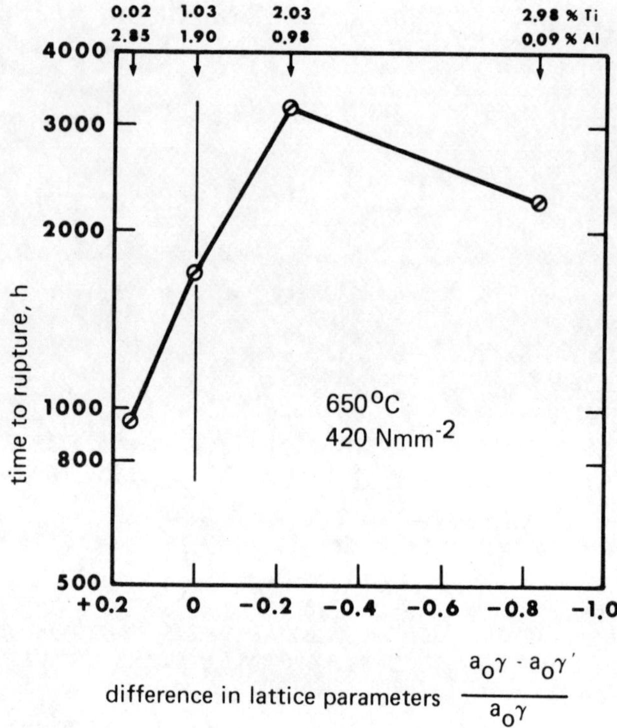

$$\frac{a_{o\gamma} - a_{o\gamma'}}{a_{o\gamma}}$$

FIG. 16.   Influence of coherency strain on creep rupture strength in superalloys.

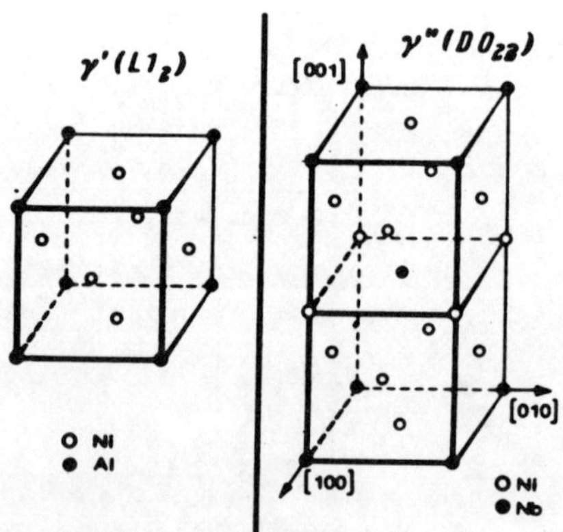

FIG. 17.   Lattice of γ'- and γ''-precipitates in superalloys.

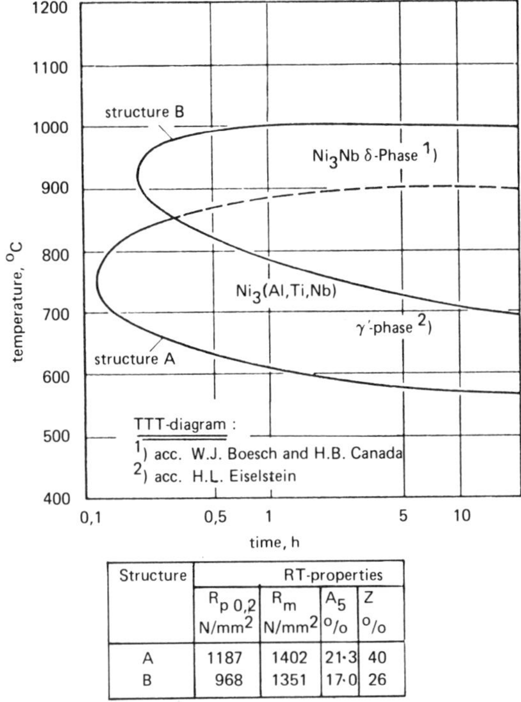

Structure	RT-properties			
	$R_{p\,0,2}$ N/mm^2	R_m N/mm^2	A_5 $^o/_o$	Z $^o/_o$
A	1187	1402	21·3	40
B	968	1351	17·0	26

FIG. 18.   TTT diagram and room temperature properties of Inconel 718.

the lattice misfit, Fig. 16,[17] the coherency strain is changed. For creep rupture strength at temperatures of about 600 °C up to 750 °C a high coherency strain is beneficial. In Table 1, two alloys with high coherency strain are given, the Fe–Ni–Cr alloys A 286 (× 5 NiCrTi 26 15) and the alloy Inconel 718. In both alloys, the γ'-phase is not stable at higher temperatures. In Inconel 718, the precipitation of fine, disc-shaped niobium-containing precipitates known as the γ''-phase of DO_{22}-type matrix, Fig. 17, is of greater importance for the 0·2% yield strength and creep rupture properties than the spherical γ' (Ni_3Al type $L1_2$) phase.[18] However at higher temperatures this γ''-phase is not stable, and needles of the δ (Ni_3Nb) orthorhombic phase arise from grain boundary tangles.[19]

While it is cooling down from the solution annealing temperature, the region of the δ-phase in the TTT diagram, Fig. 18,[20] should not be crossed, in order to avoid the occurrence of δ-precipitates at grain boundaries which may result in a loss of strength and ductility.

In the closed-die forging processes, a rounded γ'-precipitation can be used to achieve a very fine grain size (ASTM 8). Details of this 'mini-grain' process are given by J. E. Barker *et al.*[21]

The problems of the influence of heat treatment stages are only the very beginning of the discussion on the influence of long term exposure in respect of microstructural stability.

4.   INFLUENCE OF LONG TERM EXPOSURE

4.1 Austenitic High Temperature Steels and Solid Solution Strengthened Superalloys

Depending on chemical composition and thermo-mechanical history, during high temperature exposure, structural instabilities, corresponding to the kind and distribution of the carbides and the subgrain structure, occur. The TTT diagram, of an austenitic steel, AISI 316 (0·02%C–17·3%Cr–13·1%Ni–2·66%Mo–1·7%Mn–bal. Fe) and the TTT diagrams of a Co-base alloy, L605 (0·01%C–20%Cr–10%Ni–15%W–1·5Mn–bal. Co)[22] are given in Figs. 19 and 20 respectively. The carbides and topological phases like Laves, σ and χ involved in the structural changes, are given in Fig. 21. These instabilities have a strong influence on mechanical properties, Fig. 22, causing an increase in room temperature-hardness and a loss of impact strength for Inconel 617 sheets and a loss of room temperature elongation, Fig. 23, of sheet materials of Haynes Alloy 188 in comparison to Hastelloy X and two melts of L605.[23]

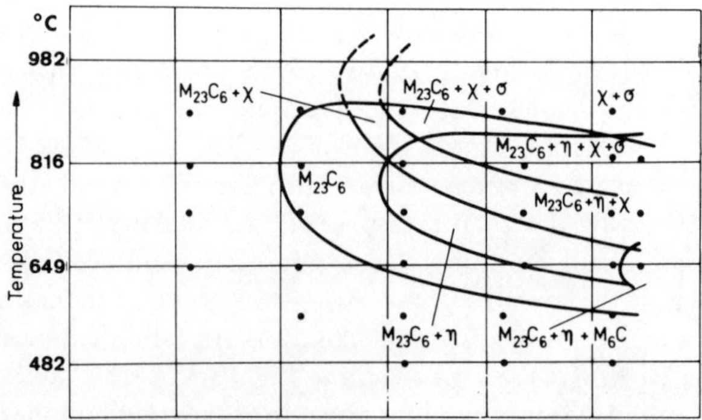

FIG. 19.   TTT diagram of an austenitic steel (AISI 316). According to Stickler.

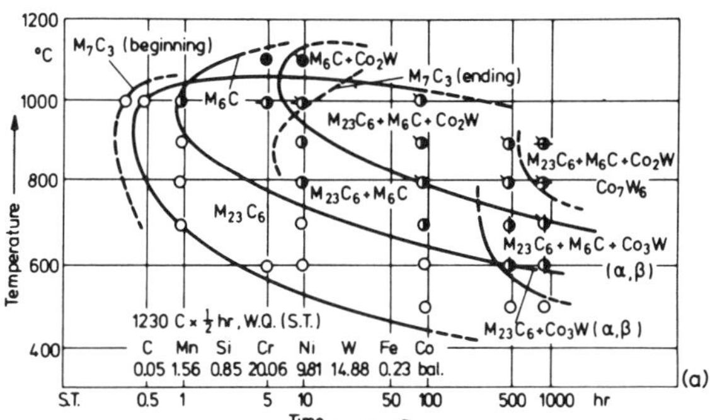

FIG. 20.   TTT diagrams for a Co-base alloy L-605 (0·01 % C–20 % Cr–10 % Ni–15 % W–1·5 Mn–bal.Co).[22]

4.2 γ'-Strengthened Material

Alloys, strengthened by γ'-precipitations with a high γ/γ' mismatch, in long term application at temperatures higher than 700 °C become prone to needle-like precipitation of a η (Ni$_3$Ti) or δ (NiNb) type. In Fig. 24, the microstructure of specimens of A 286 after long-term exposure are shown in respect to a TTT diagram. As well as the stable η-phase (hexagonal DO$_{24}$-type), with a higher silicon content, precipitation of a G-phase (chemical composition approximately (NiFe)$_{20}$/Ti$_7$Si$_8$ or, alternatively, Ni$_2$TiSi) occurs in the grain boundaries.[24]

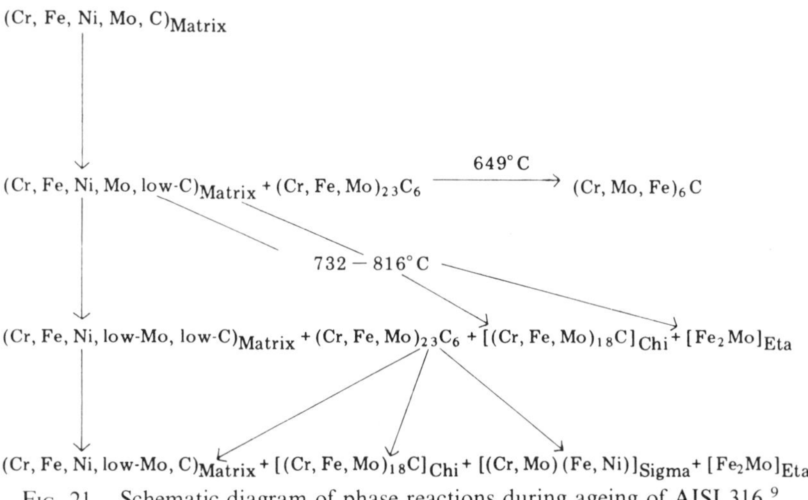

FIG. 21.   Schematic diagram of phase reactions during ageing of AISI 316.[9]

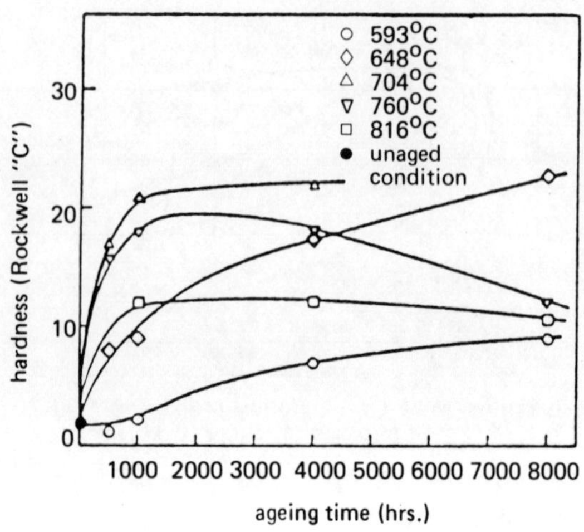

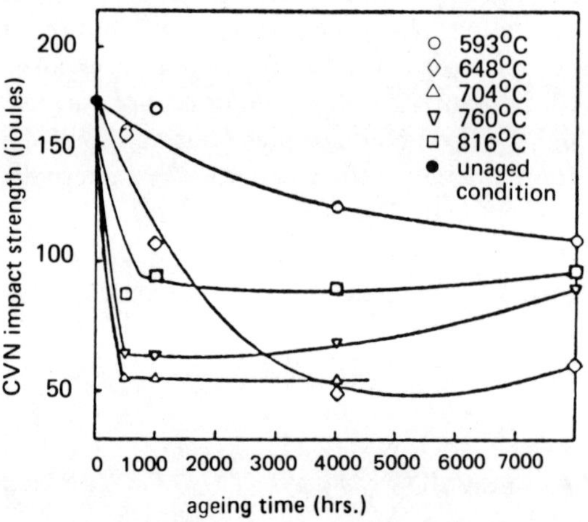

FIG. 22. Influence of long term exposure on room temperature properties of Inconel 617.

In alloys, such as Nimonic 80 A, Udimet 520 or Udimet 710, Fig. 25, with ageing at temperatures above about $0.6\,T_m$, the γ'-particles coarsen by Ostwald ripening.[25] In Fig. 26, for Udimet 520, the microstructural change is demonstrated by electron optical replica micrographs. With this ripening a change of mechanical properties, especially 0.2% yield strength, appears in aged specimens of Waspaloy and Udimet 520, Fig. 27, as expected by

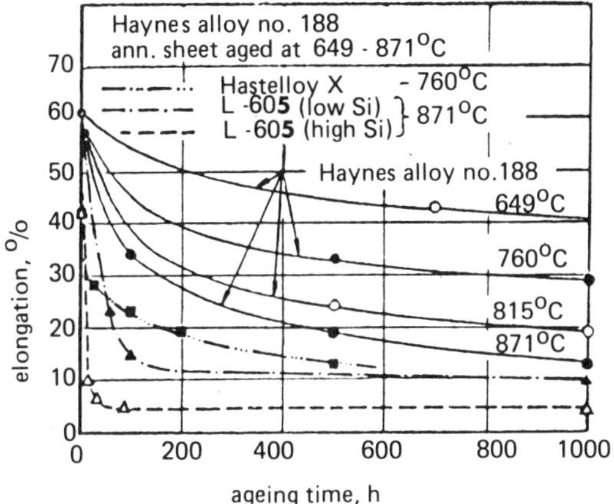

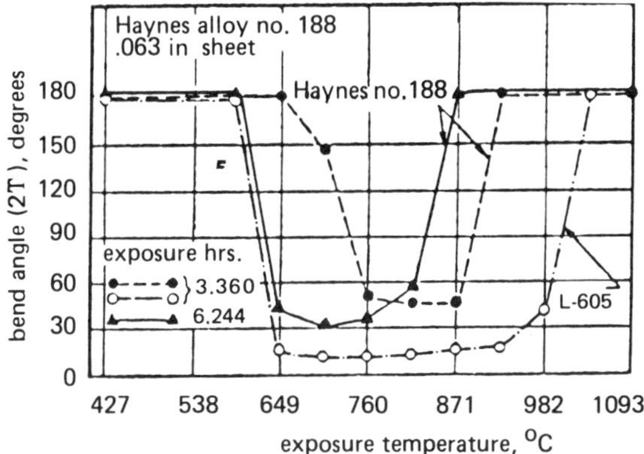

FIG. 23.   Influence of long term exposure on room temperature properties of
Hastelloy X, L 605 and Haynes Alloy 188.

theoretical approximations of the interaction of dislocations and particle
size.[25]

During creep exposure, the ripening of γ'-particles is dependent on many
structural features,[26] which will not be discussed in this paper.

It must be realised, however, that great efforts should be made to ensure
that a superalloy is cast in a N_V-number (or Phacomp) controlled manner.
In Fig. 28, the creep–rupture curve of melts of IN 738 LC, free, quasi-free
and σ-prone, are compared. The rupture strength of the σ-prone material is
less than that of the σ-free version.[27]

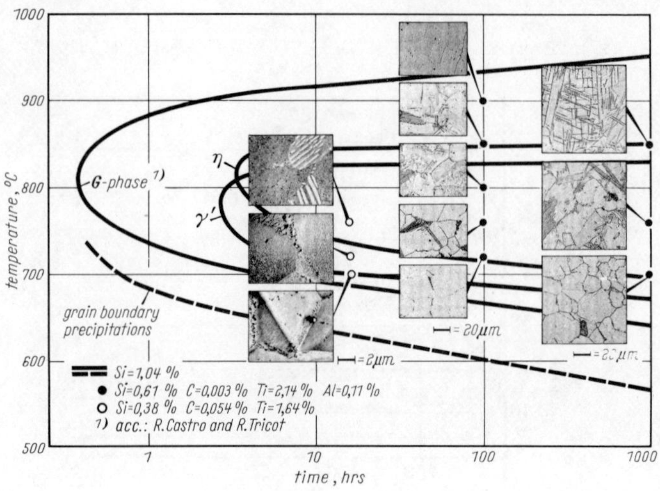

Fig. 24. Microstructures of Vacumelt ATS 28 (X-5 NiCrTi 26 15, similar to A 286).

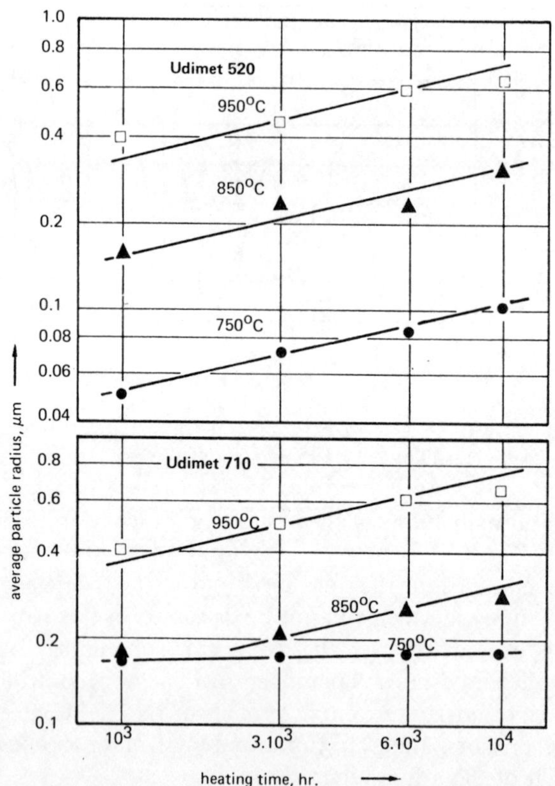

Fig. 25. γ'-Coarsening of Ni-base superalloys.

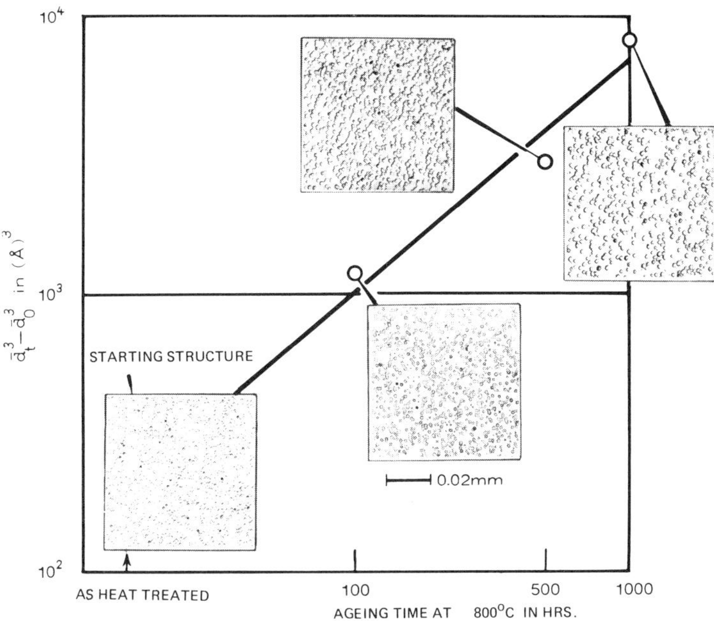

FIG. 26.   Dependence of γ'-coarsening of a Ni-base superalloy on ageing time (Udimet 520).

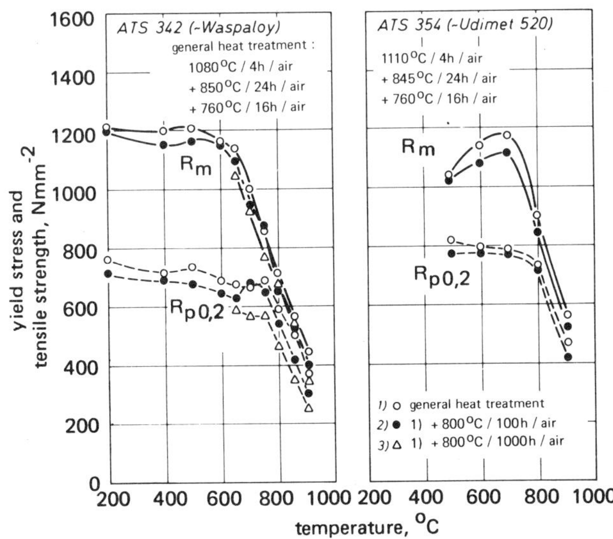

FIG. 27.   Influence of overaging on the temperature dependence of the $R_{p0.2}$ and R_m values of two Ni-base superalloys.

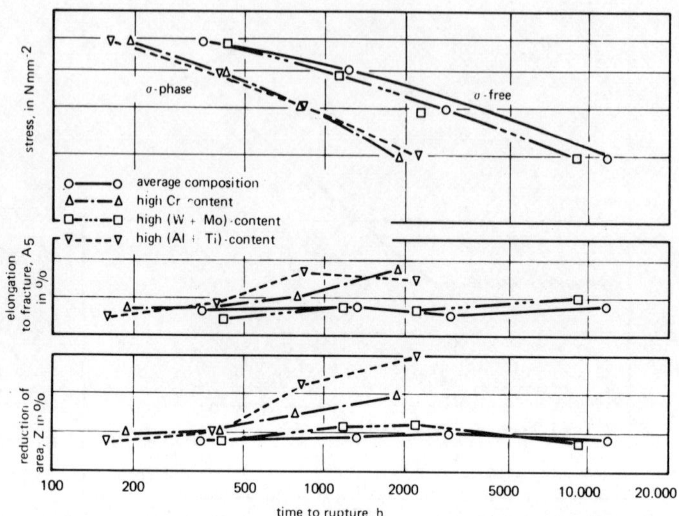

FIG. 28.   Creep–rupture properties of alloys similar to IN-738 with different N_V values.

5.   FINAL REMARKS

In this paper, some typical examples of the influence of heat treatment and long term exposure have been discussed according to simplified schemes, with the emphasis on exposure at temperature without stress. The wide field of creep behaviour related to structural properties and stability problems has not been covered. It is the task of other papers to introduce these problems.[13,28,29,30,31]

REFERENCES

1. (1972). *The Superalloys* (ed. C. T. Sims and W. C. Hagel), John Wiley & Sons, New York.
2. Proceedings of the 1st Int. Symposium: 'Structural Stability in Superalloys', Seven Springs, USA, 1968.
3. Proceedings of the 2nd Int. Symposium: 'Superalloys Processing', Seven Springs, USA, 1972.
4. Proceedings of the 3rd Int. Symposium: 'Superalloys—Metallurgy and Manufacture', Seven Springs, USA, 1976.
5. Bongartz, K., von den Steinen, A. and Schubert, F. (1972). *Zeitschrift für Werkstofftechnik*, **3**(4), 176–84.

6. (1974). *High Temperature Materials in Gas Turbines, Proceedings of a Symposium at Baden, March 1973* (ed. P. Sahm and M. O. Speidel), Elsevier Scientific Publishing Company, Amsterdam.
7. (1978). *High Temperature Alloys for Gas Turbines*, Proceedings of COST 50-Meeting (ed. D. Coutsouradis, P. Felex, H. Fischmeister, L. Habraken, Y. Lindblom and M. O. Speidel), Applied Science Publishers, London.
8. von den Steinen, A. 'Hochwärmfeste Stähle', Proceedings of Kontaktstudium Werkstoffe Eisen und Stahl IV: Festigkeits- und Bruchverhalten bei höheren Temperaturen. Schubert, F. 'Hochwärmfeste Legierungen', *ibid.* published by VDEH, Verlag Stahleisen MBM, Düsseldorf.
9. Weis, B. and Stickler, R. (1972). *Met. Trans.*, **3**, 851.
10. Orr, J. (1978). 'A Review of the Structural Characteristics of Alloy 800', Alloy 800: Proceedings of the Petten International Conference.
11. Schubert, F. and Horn, H. (1974). *Thyssen Techn. Ber.*, **6**(1), 43/52.
12. Moon, D. M. and Stickler, R. (1974). *Phase Stability in Superalloys, Proceedings of a Symposium at Baden, March 1973*, (ed. P. Sahm and M. O. Speidel), Elsevier Scientific Publishing Company, Amsterdam.
13. Schubert, F. *et al.* (1975). 'Entwicklungsgrundsätze für die Verfestigung von hochwarmfesten Werkstoffen durch ausgeschiedene und dispergierte Teilchen' aus DGM-Symposium Warmformgegung und Warmfestigkeit.
14. Schubert, F. and von den Steinen, A. (1976). *TEW-Techn. Berichte*, **2**, (6H2), 172.
15. Schubert, F. (1978). Feinguss von Turbinenschaufeln hoher Temperaturfestigkeit und Langer Standzeit at Werkstofftechnische Probleme bei Gasturbinentriebwerken, (ed. H. Hansen, P. Esslinger), Werkstofftechnische Verlagsgesellschaft, Karlsruhe.
16. Loomis, W. (1972). *Met. Trans.*, **3**, 988.
17. Manier, G. N. and Bridge, J. E. (1971). *Met. Trans.*, **2**, 95, 102.
18. Oblak, J. M., Paulomis, D. E. and Duvall, D. S. (1974). *Met. Trans.*, **5**, 143–53.
19. Moll, J. H., Maniar, G. N. and Muzyka, D. R. (1971). *Met. Trans.*, **2**, 2153–71.
20. Schubert, F. and Horn, H. (1974). *Thyssen Techn. Ber.*, **1**, 43–52.
21. Barker, J. E., Ross, E. W. and Radawich, J. F. (1970). *J. Metals*, **1**, 31–41.
22. Takeda, S. and Yukawa, N. (1967). *Nippon Kinzoku Grakkai-ho*, 6, 11, 12, 784, 850 (cited by Stickler in reference 6).
23. Haynes Alloy 188: *Aerospace Structural Metals Handbook*, Code 4310.
24. Castro, R. and Tricot, R. (1964). *Mem. Sci. Rev. Metallurg.*, **61**, 573.
25. Susukida, H., Sakumoto, Y., Isuji, I. and Kawai, H. (1972). 'Strength and Microstructure of Ni-base Superalloys after Long Term Heating', Kober Technical Institute, Akashi, Japan.
26. Schubert, F. (1972). *DEW-Techn. Ber.*, **II**, 2, 94.
27. Schubert, F. (1977). Abschlußbericht zu COST 50, BCT 34, D 2/1.
28. Brandis, H., Horn, E. and Schubert, F. (1976). *Archiv für Eisenhüttenwesen*, **47**, (L), 113.
29. Kear, B. H., Leverant, G. R. and Oblak, J. M. (1969). *Trans. ASM*, **62**, 639.
30. Coply, S. M. and Kear, B. H. (1967). *Trans Methargic. Soc. AIME*, **293**, 977.
31. Kear, B. H., Giamei, A. F., Leverant, G. R. and Oblak, J. M. (1969). *Scripta Metallurgica*, **3**, 123, 445; (1970) **4**, 567.

Index